Leigh Hunt

Table Talk

To which are added imaginary conversations of Pope and Swift. Vol. 1

Leigh Hunt

Table Talk
To which are added imaginary conversations of Pope and Swift. Vol. 1

ISBN/EAN: 9783337240196

Printed in Europe, USA, Canada, Australia, Japan

Cover: Foto ©Lupo / pixelio.de

More available books at **www.hansebooks.com**

APPLETONS' NEW HANDY-VOLUME SERIES.

TABLE-TALK.

TO WHICH ARE ADDED

IMAGINARY CONVERSATIONS OF POPE AND SWIFT.

BY

LEIGH HUNT.

NEW YORK:
D. APPLETON AND COMPANY,
549 AND 551 BROADWAY.
1879.

PREFACE TO ORIGINAL EDITION.

THE title of this volume, "TABLE-TALK," will, it is hoped, be found by the reader to be warranted by the conversational turn of the style, as well as the nature and variety of the subjects touched upon, and the manner in which they are treated. Some portion was really talked; and it may be said of the rest, that the thoughts have, in all probability, passed the writer's lips in conversation.

The "Imaginary Conversations of Pope and Swift" were considered an appropriate addition to a volume of "Table-Talk," and are intended strictly to represent both the turn of style and of thinking of these two poets; though the thoughts actually expressed are the writer's invention.

On correcting the sheets for press, I am not aware of any remark that I should particularly wish to modify, with the exception of something that is said of Germany in the course of the article on "Goethe." I have since become better acquainted with the great intellects of that nation, which has unquestionably produced the leading thinkers of the century. The world has yet to learn the extent of its obligations to such men as Goethe and Schiller, to Lessing, to Kant, to Herder, Richter, Fichte, and others.

LEIGH HUNT.

CONTENTS.

	PAGE
Table-Talk	11
Ladies carving at Dinner	13
Anomalies of Dishes and Furniture, etc.	14
Topics for Dinner	15
Wild Flowers, Furze, and Wimbledon	17
Mistakes of the Press	21
May-Time	22
Malice of Fortune	25
Bishops and Brahmins	25
The "Blessed Restoration"	30
The Sun	32
Bon-mot of a Coachman	32
Song of the Nightingale	34
Ovid	35
The Voice of the Rook	35
How Lawyers go to Heaven	36
Collins, the Poet	36
A Fact	40
The Two Conquerors	41
Clerical Titles	41
Horace Walpole and Pinkerton	42
Jews	44

CONTENTS.

	PAGE
SMOLLETT	45
CHEMISTRY	47
PETTY CONVENIENCES AND COMFORTS	49
TEARS	50
DR. ALDRICH	51
LORD MARCHMONT'S RECEIPT FOR LONGEVITY	52
THE AMERICAN REVOLUTION	54
DISCOVERERS OF AMERICA	54
WONDER NEVER CEASES	55
DALY, THE DUBLIN MANAGER	56
LIGHT AND COLORS	59
VERSIONS OF ANCIENT LYRICS	61
CATHARINE II. OF RUSSIA	62
PETRARCH AND LAURA	63
MORAL AND PERSONAL COURAGE	65
TIGHT-LACING	66
GRAVITY AND INDUSTRY OF DANCERS	68
ADVERTISEMENTS	70
SPORTSMEN AND CUSTOM	70
BEARS AND THEIR HUNTERS	71
SELF-STULTIFICATION	73
COWSLIPS	74
APRIL FOOLS	75
PRIVATE WAR	77
BEAUMARCHAIS	80
MOZART	81
VIOLET—WITH A DIFFERENCE	81
VERBAL MISTAKES OF FOREIGNERS	81
HUME AND THE THREE LITTLE KINGS	83
A CHARMING CREATURE	84
BACON	84
SUICIDES OF BUTLERS	84
DUELS	85
LISTON	87

CONTENTS.

	PAGE
STEEPLE-CHASING	88
TURKEYS	89
BAGPIPES	90
CÆSAR AND BONAPARTE	90
PSEUDO-CHRISTIANITY	91
DYED HAIR	93
EATING	95
POLAND AND KOSCIUSKO	99
ENGLAND AND THE POPE (GREGORY)	100
THE DUKE OF WELLINGTON'S CONCERT	101
WAR, DINNER, AND THANKSGIVING	104
FIRES AND MARTYRDOM	106
RESPECTABILITY	108
USE OF THE WORD "ANGEL," ETC., IN LOVE-MAKING	113
ELOQUENCE OF OMISSION	113
GODS OF HOMER AND LUCRETIUS	114
AN INVISIBLE RELIC	115
A NATURAL MISTAKE	115
MORTAL GOOD EFFECTS OF MATRIMONY	115
UMBRELLAS	115
BOOKSELLERS' DEVICES	117
WOMEN ON THE RIGHT SIDE	118
SHENSTONE MISTAKEN	118
THE MARSEILLES HYMN	119
NON-SEQUITUR	120
NON-RHYMES	120
STOTHARD	121
THE COUNTENANCE AFTER DEATH	124
HUME	124
GIBBON	124
ANGELS AND FLOWERS	125
AN ENVIABLE DISTRESS	126
SIR THOMAS DYOT	126
ANCIENT AND MODERN EXAMPLE	127

CONTENTS.

	PAGE
Milton and his Portraits	130
William Hay	131
Bishop Corbet	132
Hoadly	133
Voltaire	133
Handel	133
Montaigne	135
Waller	136
Otway	137
Raphael and Michael Angelo	137
Wax and Honey	138
Associations with Shakespeare	139
Bad Great Men	140
Cicero	140
Flowers in Winter	140
Charles Lamb	141
Sporting	143
Wisdom of the Head and of the Heart	147
Mæcenas	147
Lord Shaftesbury's Experience of Matrimony	148
A Philosopher thrown from his Horse	148
Worlds of Different People	149
Mrs. Siddons	149
Non-Necessity of Good Words to Music	151
Goethe	152
Bacon and James I.	156
Goldsmith's Life of Beau Nash	156
Julius Cæsar	157
Fénelon	158
Spenser and the Month of August	158
Advice	160
Eclipses, Human Beings, and the Lower Creation	161
Easter-Day and the Sun, and English Poetry	163
The Five-Pound Note and the Gentleman	165

	PAGE
PAISIELLO	167
CARDINAL ALBERONI	168
SIR WILLIAM PETTY THE STATIST AND MECHANICAL PHILOSOPHER	169
NAME OF LINNÆUS	170
JOHN BUNCLE (THE HERO OF THE BOOK SO CALLED)	170
POUSSIN	171
PRIOR	172
BURKE AND PAINE	173
THE DUTCH AT THE CAPE	175
RUSSIAN-HORN BAND	175
DOGS AND THEIR MASTERS	176
BODY AND MIND	178
WANT OF IMAGINATION IN THE COMFORTABLE	179
THE SINGING MAN KEPT BY THE BIRDS	182
A STRANGE HEAVEN	184
STANDING GODFATHER	186
MAGNIFYING TRIFLES	187
RELICS	187
SOLITUDE	188
LOUIS XIV. AND GEORGE IV.	188
HENRY IV. OF FRANCE AND ALFRED	189
FELLOWS OF COLLEGES	189
BEAUTY A JOY IN HEAVEN	190
ASSOCIATIONS OF GLASTONBURY	190
LIBERTY OF SPEECH	190
WRITING POETRY	191
THE WOMEN OF ITALY	191
FRENCH PEOPLE	192
THE BLIND	192
LONDON	192
SOUTHEY'S POETRY	193
VULGAR CALUMNY	193
VALUE OF ACQUIREMENTS	194

	PAGE
THE BEARD	195
ATTRACTIONS OF HAM	195
SLEEPING UNDER THE SKY	196
WAR POETRY	197
MONEY-GETTING	198
VALUE OF WORDS	198
UNWRITTEN REVELATIONS	198
WEEPING	199

IMAGINARY CONVERSATIONS OF POPE AND SWIFT.

CONVERSATION OF POPE	203
CONVERSATION OF SWIFT AND POPE	221

TABLE-TALK.

TABLE-TALK

Is so natural to man, that the mouth is the organ both of eating and speaking. The tongue is set flowing by the bottle. Johnson talked best when he dined; Addison could not talk at all till he drank. Table and conversation interchange their metaphors. We *devour* wit and argument, and *discuss* a turkey and chine. That man must be very much absorbed in reflection, or stupid, or sulky, or unhappy, or a mere hog at his trough, who is not moved to say something when he dines. The two men who lived with no other companions in the Eddystone Lighthouse, and who would not speak to one another during their six months, must have been hard put to it when they tapped a fresh barrel. To be sure, the greater the temptation the greater the sulk; but the better-natured of the two must have found it a severe struggle on a very fine or very foggy day.

Table-talk, to be perfect, should be sincere

without bigotry, differing without discord, sometimes grave, always agreeable, touching on deep points, dwelling most on seasonable ones, and letting everybody speak and be heard. During the wine after dinner, if the door of the room be opened, there sometimes comes bursting up the drawing-room stairs a noise like that of a tap-room. Everybody is shouting in order to make himself audible; argument is tempted to confound itself with loudness; and there is not one conversation going forward, but six, or a score. This is better than formality and want of spirits; but it is no more the right thing than a scramble is a dance, or the tap-room chorus a quartet of Rossini. The perfection of conversational intercourse is when the breeding of high life is animated by the fervor of genius.

Nevertheless, the man who can not be loud, or even vociferous on occasion, is wanting on the jovial side of good-fellowship. Chesterfield, with all his sense and agreeableness, was but a solemn fop when he triumphantly asked whether anybody had "ever seen him laugh?" It was as bad as the jealous lover in the play who says: "Have *I* been the life of the company? Have *I* made you all die with merriment?" And there were occasions, no doubt, when Chesterfield might have been answered as the lover was: "No; to do you justice, you have been confoundedly stupid."

Luckily for table-talkers in general, they need be neither such fine gentlemen as Chesterfield, nor such oracles as Johnson, nor such wits as Addison and Swift, provided they have nature and sociability, and are not destitute of reading and observation.

LADIES CARVING AT DINNER.

Why doesn't some leader of the fashionable world put an end to this barbarous custom? What a sight, to see a delicate little creature, or, worse perhaps, a "fine woman," in all the glory of her beauty and bedizenment, rise up with a huge knife in her hand, as if she were going to act the part of Judith, and begin heaving away at a great piece of beef! For the husband does not always think it necessary to take the more laborious dish on himself. Sometimes the lady grows as hot and flustered as the housewife in the "Winter's Tale," "her face *o' fire* with labor." Gentlemen feel bound to offer their services, and become her substitutes in that unseemly warfare. Why don't they take the business on themselves at once? or, rather, why don't they give it to the servants, who have nothing better to do, and who have eaten their own meal in comfort? A side-table is the proper place for carving. Indeed, it is used for that purpose in some great houses. Why not in all? It is favorable for additional means of keeping the dishes hot; nobody at the dinner-table is inconvenienced; and the lady of the house is not made a spectacle of, and a subject for ridiculous condolements. None would regret the reformation but epicures who keep on the watch for tidbits, to the disadvantage of honest diners, and whom it would be a pleasure to see reduced from shocking oglers at the hostess into dependents on the plebeian carver.

ANOMALIES OF DISHES AND FURNITURE, Etc.

Among the customs at table which deserve to be abolished is that of serving up dishes that retain a look of "life in death"—codfish with their staring eyes, hares with their hollow countenances, etc. It is in bad taste, an incongruity, an anomaly; to say nothing of its effect on morbid imaginations.

Even furniture would be better without such inconsistencies. Claws, and hands, and human heads are not suited to the dead wood of goods and chattels. A chair should not seem as if it could walk off with us; nor a table look like a monstrous three-footed animal, with a great flat circular back, and no head. It is such furniture as the devil might have had in Pandemonium—

"Gorgons, and hydras, and chimeras dire."

A lady sometimes makes tea out of a serpent's mouth; and a dragon serves her for a seat in a garden. This is making a witch of her, instead of a Venus or a Flora. Titania did not sit on a toadstool, but on a bank full of wild thyme and violets.

This bad taste is never more remarkably exemplified than in the case of fountains. The world seems to have given fountains a privilege for exciting incongruous and filthy ideas; for nobody, as far as I am aware (except Pope, by an implication), has protested against their impossible combinations and vomiting mouths; than which nothing surely can be more ridiculous or revolting. A fountain should suggest nothing but feelings of purity and freshness; yet they go to the reverse extreme, and seem to endeavor to make one sick.

TOPICS FOR DINNER.

What a thing it is to sit down to dinner, after reading of the miseries in starving countries! One fancies one has no right to eat and drink. But the thought must be diverted; not because the question is to be got rid of on every other occasion—quite the contrary; but because, having done his best for it, great or small, then, and in that case only, the conscientious diner has a right to waive it. Dinner is a refreshment, and should be such, if possible, to everybody, and most of all to the anxious. Hence the topics fittest for table are such as are cheerful, to help digestion; and cordial, to keep people in heart with their fellow creatures. Lively anecdotes are of this description—good-humored personal reminiscences, literary chat, questions as easy to crack as the nuts, quotations flowing as the wine, thoughts of eyes and cheeks blooming as the fruit, and beautiful as those that have looked at us over the mutual glass. They poet says:

> "What, and how great, the virtue of the art
> To live on little with a cheerful heart,
> Let's talk, my friends, but talk before we dine!"

Yes, but not even then, *just* before we dine. A man's in a very bad disposition for living on little before he dines. He is much more disposed to do so afterward, particularly if he has eaten too much. The time for discussing anxious subjects, especially those that regard the poor, is neither at dinner, when the topic becomes almost indecent; nor just before it, when hunger is selfish; nor just after it, when the feelings are too self-complacent; but at moments when the pulse is lowered, without being too much so for reason; though, indeed, if legislators could be kept without their dinners for some two or three days, there are occasions when people might be the better for it. Members of Parliament hardly see fair play between their dinner-bell and the calls of the many; and, when the wine is in, the perfection of *wittenagemot* wisdom is apt to be out. The prince in Voltaire thought his people happy "when he had dined."

" Quand il avait dîné, croyoit son peuple heureux."

Luckily, we have princes, and a Parliament too (whatever be its faults), that can dine happily, and yet not believe typhus and famine comfortable.

WILD FLOWERS, FURZE, AND WIMBLEDON.

Those flowers on the table are all wild flowers, brought out of ditches, and woodsides, and the common; daisies and buttercups, ground-ivy, hyacinths, violets, *furze:* they are nothing better. Will all the wit of man make anything like them?

A. Yes, paintings.

B. And poetry and music.

C. True; but paintings can not be sown; they can not come up again every spring, fresh and fresh, beautiful as ever.

A. Paintings are sown by copyists and engravers.

C. Very true indeed; but still there is a difference. Humphreys is not Correggio; Linton is not Rembrandt; Strange himself is not Titian. The immortal painter does not survive in person to make even his own reds and blues immortal as his name. Yet here is the hyacinth, as fresh as when it was first created. Here is Burns's

"Wee, modest, crimson-tipped flower,"

as new as if the great peasant had just turned it up with his plow.

B. Poetry seems as if it would last as long as flowers ; and it has no need of renewal.

C. God forbid I should undervalue his most wonderful work here on earth, the creature who can himself create ! I wonder what they have to resemble, or surpass him, in the planets Mercury and Venus ? I suppose he gets better and better as he nears the sun ; and in the sun is the heaven we are all going to ; not the final heaven, but just a kind of celestial half-way house ; our own earth made heavenly after a human fashion, to enable us to take by degrees to beatitude.

B. There have been worse fancies about the sun than that.

D. Don't condescend to mention 'em.* The very best must be unworthy of the orb whose heat and light are the instruments for making all these beautiful things. And yet, unless you would have everything there lilies and roses, can you conceive any covering fitter for the hills of the sun

---

* Nothing is meant here to be insinuated against speculations like those of the "Vestiges"; compared with which, nine tenths of all the theology that was ever theologized are but so much ignorant and often impious babble.

itself than this magnificent furze, as it now appears here in England, robing our heaths and commons all over the country?

There is an advertisement in the papers announcing a building project at Wimbledon and Westhill. The houses are to occupy a portion of Wimbledon Park; and boards are put among the trees by the roadside, boasting of the "fine frontage." Well may they boast of it, especially at this season of the year. It is a golden undulation; a foreground, and from some points of view a middle distance, fit to make the richest painter despair; a veritable Field of Cloth of Gold. Morning (Aurora, the golden goddess), when the dawn is of a fineness to match, must look beauty for beauty on it. Sunset is divine. The gold goes stretching away in the distance toward the dark trees, like the rich evening of a poetic life. No wonder Linnæus, when he came to England and first beheld this glorious shrub in bloom, fell down on his knees, and thanked God that he had lived to see it. No wonder statesmen and politicians go forth to lodge about the place for a little while, to procure air and refreshment; perhaps to get a new lease of existence; perhaps to die where they may still find something beautiful on earth—beautiful enough to comfort their mistakes about it, and to prepare them for a place where it is easier to look for flowers than revolutions. As to figures in the landscape, they are not many, nor discordant; such as a horse or two, a few cattle, now and then a horseman, or a sturdy peasant on foot, or a beauty in a barouche. Sometimes the peasant is aged, but hale; or sturdy, though but a child;—signs both of good air, and prosperity, and a true country spot. I hardly know which is the more picturesque sight—a fine, ruddy-cheeked little peasant-boy, not beyond childhood, coming along with a wheelbarrow full of this golden furze, his face looking like a bud a-top of it; or a bent, hearty old man (bent with age, not with his perquisite) carrying off a bunch of it on his back, as if he triumphed over time and youth.

Sometimes you meet a lady coming with a bunch of hyacinths; sometimes a fine young fellow of a gentleman, who has not disdained to stick a bit of furze in his coat. It is not the love of flowers that makes people effeminate, but indoor habits that produce a craving for stimulants and dread of trouble. This very Wimbledon Park was once occupied by a cultivator, and even painter of flowers, whom nobody that didn't know him, and beheld at his gentle tasks, would have suspected to have been General Lambert, one of the boldest and most independent of the officers of Cromwell. He lived there in the interval between his rival's elevation to sovereign power and the return of Charles II., and was famous for the sums he gave for his pinks and tulips.

MISTAKES OF THE PRESS.

The annals of law and typography contain the remarkable fact that an edition of the Bible was once printed, in which the word *not*, to the horror and consternation of the religious world, was left out of the seventh commandment! They called for its restoration with an impatience more creditable to their zeal than their sense of security; while, on the other hand, some daring theologians (who, like the Catholics, did not think themselves tied in every respect to those letters of the old law) doubted whether, for the sake of the commandment itself, the omission had not better remain as it was, seeing that, "in nine cases out of ten, the prohibition was the temptation."

Mistakes of the press have given rise to such ludicrous combinations, that a small wit (Caleb Whitford) obtained a reputation solely by a few articles about them in a newspaper. I never, in the course of my own experience, met with one of a more astounding aspect than the following. It is innocent of all scandal, or libel, or double meaning. It was a pure mistake of the printer, ludicrously unintelligible, and threw the readers into agonies of conjecture. The writer had observed that, "although there is no mention either of coffee or tobacco in the 'Arabian Nights,' the former, from association of ideas with existing Eastern manners, always reminded him of that delightful book"; and then followed this extraordinary sentence: "*as sucking does for the snow season.*"

This mistake was so high, abundant, and ridiculous, that, if I remember rightly (for the article was my own), I refused to correct it. I thought it better to leave it as it stood, for a perpetual pleasure of astonishment to all who might chance to light upon the pages in which it occurred.

The proper words, however, were these: "as *smoking* does for the *same reason.*"

MAY-TIME.

Such a delightful commencement as we have had of the month of May is a perfect godsend; for our climate is seldom so lucky. May is a pretty word; a charming thing in books and poets; beautiful always in some degree to look at, as far as hedges and trees go, whatever be the state of the weather; that is to say, provided you can quit the fireside, and the windows are not too misty with rain to see through. But the hedges in general succeed better than the skies. There is apt to be more blossom than sunshine; and people lie in bed on May-morning, and wonder what possessed their ancestors, to induce them to get up at dawn, and go poking about the wet bushes.

I suspect it was never very easy to reckon

upon a fine May-day in England. If the wind was in a good quarter, the chances were that it rained; and, if the sky was clear, then probably the wind was in the east.

"Rough winds do shake the *darling buds* of May,"

says a lovely verse in Shakespeare. Our ancestors, however, had more out-of-door habits than we, and seem to have cared little for east winds. You hear a great deal more of north winds than east in the old writers. At the same time, we must not forget that our May-day comes nearly a fortnight sooner with us than it did with them. The change took place when the calendar was altered, about a hundred years back; and the consequence was, that the May-day of our ancestors now falls on the 12th of the month. The circumstance gave rise to some verses by Mr. Lovibond, a gentleman "about town" in the days of Chesterfield and Walpole, which the subject (and the prevailing bad taste in verses) rendered popular. They were called "The Tears of Old May-day." This is the way in which Mr. Lovibond laments:

"Onward *in conscious majesty* she came,"

(To wit, poor May!)—

"The grateful honors of mankind to *taste*,
To gather fairest wreaths of *future fame*,"

(What is the meaning of that?)

"And blend fresh triumphs with her glories past.

"Vain hope! No more in choral bands unite
　Her virgin vot'ries; and at early dawn,
Sacred to May and Love's mysterious rite,
　Brush the light dew-drops from the spangled lawn.

"To her no more Augusta's *wealthy pride*
　Pours the full tribute from Potosi's mine!!
Nor fresh-blown garlands village maids provide,
　A purer offspring at her rustic shrine," etc., etc.

What does the reader take to have been "the full tribute from Potosi's mine"? It was the plate which the milkmaids used to borrow to decorate their Maypole.

Compare with this stuff the fresh, impulsive verses and bright painting of Spenser:

"Then came *fair May, the fairest maid* on ground,
Decked all with *dainties* of her season's pride,
And throwing flowers out of her lap around.
Upon two brethren's shoulders she did ride,
The twins of Leda: which, on either side,
Supported her, like to their sovereign queen.
Lord! how all creatures laughed when her they spied;
And leaped and danced as they had ravished been;
And Cupid's self about her fluttered all in green."

If people, then, have a mind to try the proper old May-day, and be up and out of doors among the blossoms when Shakespeare was, or Spenser's Rosalind, or the pretty queen of Edward IV. (for royalty used to go a-Maying once), next

Tuesday is their time, supposing the weather favorable, and good folks " in a concatenation accordingly." Only they must take care how they are too merry ; otherwise, they will wake the Tractarian old lady next door, who will think the world is going to be at an end if people are not as sleepy and stupid as herself.

MALICE OF FORTUNE.

Mr. Green, the aëronaut, has had an escape from a death which would have looked like a mockery. He was near being killed by his balloon, not aloft in the clouds, or by a descent like Phaëton, but in a *cart* in which he was riding upon it, like the Welshman on his cheese in the "Splendid Shilling." Mr. Green's courage is to be congratulated on not having brought him to so mock-heroical a pass. The greatest trick of this sort ever played by Fortune was the end of Bruce the traveler, who, after all his perils by flood and by field, from wars, from wild beasts, from deserts, from savage nations, broke his neck down his own staircase at home ! It was owing to a slip of the foot, while seeing some visitors out whom he had been entertaining.

This was the very anti-climax of adventure.

BISHOPS AND BRAHMINS.

I hold the Church of England in great respect for several reasons. One is, that it lets me hold my own form of Christian opinion without molestation; another, that having reformed itself once, and to no little extent, it can do so again, I have no doubt, and would to-morrow if it had its free way, and so give the *coup de grâce* in this country to the last pretenses of Popery. A third reason is, that its clergy, upon the whole, and, considering their number, are the best behaved, most learned and most reasonable, most gentle, most truly Christian, in Europe; the occasional excesses of individuals among them, however enormous, being far less than the crimes and catastrophes of those in Catholic nations, originating in causes which need not be dwelt upon.

But the reasonableness and well-tempered security of ordinary clerical existence in this country give rise in some instances to scandals, injurious in proportion to their very seeming warrant.

Why do bishops, who won't go to theatres, accept invitations to public dinners? They had much better be seen at the representation of "Lear" or "Macbeth" than at a Lord Mayor's feast. It has an unseemly look at any time, especially in your fat bishop, and most especially when the reports of the feast in the newspapers are followed by accounts of the starving poor. If such tremendous inequalities in the social condition are not to be remedied, why mortify the sufferers? And if they are, why exasperate them? Reports of public dinners, let the guests be who they may, harmonize ill with those of the police-office and the Poor-Laws; but, when bishops are among the diners, the scandal is doubled, and one is astonished they do not see it. But a bishop seems to see nothing else, when a dish is before him. Observe—the world would have no objection to his being fat and jolly, if he made no saintly pretensions, or if he could square it with appearances in other respects, and his duties to the unfed. There is F., who is as fat as any one of them, and who has brains and activity enough for the whole bench. If they could all bestir themselves in behalf of the poor as he does, and manifest as unclouded an intellect, I am not sure the public would not rejoice in their obesity, and regard it as the right and privilege of those who endeavored to spread a table for mankind. Who could have grudged his fat to Berkeley? or to Luther? or to good Bishop Jewel (if he had it)? or to that pattern of a prelate, who thought it a shame to have a hundred pounds in the hands of his steward? But when bishops and their families grow rich, while the poor grow poorer, and when it is the rarest thing in the world (with the exception, now and then, of a Thirlwall or a Stanley) to find them attending a public meeting but for selfish or corporate purposes, people naturally dislike to see them fat and feeding, especially when they come in the lump together, as at these Lord Mayor's feasts. Bishops should never appear in flocks, like vultures. There is an advertiser of after-dinner pills, who recommends the drug by long lists of his patrons, including almost the whole of the right reverend bench. The sight is laughable, to say the least of it. Many honest friends of the Establishment think it deplorable. It is a positive proclamation of excess; an ostentation of apoplexy; a telling the world, that to be a bishop and to want boxes of pills is the same thing. Or, if we are to take it as a mere matter of indifference and nonchalance, it becomes so much the worse.

ADVERTISER (*asking permission to boast of his "favors"*). "My Lord Bishop, may I tell the world what good my pills do to your lordship's indigestion?"

BISHOP. "Oh, certainly."

The Hindoo gentry have a custom among them of giving feeds to their bishops, the Brahmins. It is a fashion—an emulation—and practiced on great family occasions. Every nobleman tries how he can outdo the rest of his class in the number of reverend personages he can get together, and the amount of food he can induce them to swallow. If only six Brahmins are brought to the verge of apoplexy, he thinks himself ruined in the eyes of his neighbors. What will the world say if there is no sickness? How can he hold up his head should no clergyman be carried away senseless? Accordingly, toward the end of the entertainment, the host may be seen (this is no fiction) literally beseeching their lordships the Brahmins to get down another plate of curry.

"I've eaten fourteen," says one of them, gasping.

"And I fifteen," says another.

"For God's sake," says the host.

"Impossible," says the Brahmin.

"But consider, my dear lord, you ate seventeen at Ram Bulkee's."

"You are misinformed, my dear sir."

"Pardon me, they were counted to his immortal honor."

"Thirteen only, on—my—sacred word."

"Don't favor me less, I implore you. See—only this one other mouthful."

"Impracticable."

"I've rolled it up, to render it the more easy."

"Consider my jaws."

"But, dear lord—"

"Have pity on my œsophagus."

"But my name, my name—"

"My—dear—son, stomachs have their limits."

"But not your lordship's generosity."

WIFE (*interposing*). "It will be the death of my husband, if you don't oblige him."

"Well, this one—(*swallowing*). Ah—my—dear—son!—(*Aside to himself.*) Why did our caste establish this custom? It might have been salutary once ; but now—O Ram! Ram! I can bear it no longer."

One other mouthful, however, still is got down, the host is a man of such meritorious wealth ; yet he was obliged to implore it with tears in his eyes. The Brahmins in vain pointed to their own. The host, with inexorable pathos, entreats them to consider the feelings of his wife and children. The mouthful is achieved, Ram Bulkee beaten, and the reverend feasters are carried off to bed, very nearly victimized by "the wisdom of their ancestors" and clarified butter.

Such are the inconveniences that may arise from customs of our own contriving ; and such the corporate resemblances among the priesthood of the most distant countries, which Christian bishops might do well to avoid.

THE "BLESSED RESTORATION."

The public are beginning to show symptoms of dislike to the anniversary of what is equivocally called the *Blessed Restoration*, and the retention of it in so grave a place as the church. The objection is not new ; but it comes with new force at a time when some antics of superstition have induced the growing intelligence of the community to look at the abuses of religion in general and to wish to see it freed from every species of scandal. People have certainly been in the habit of taking strange occasions for expressing their gratitude to Heaven; and this "Blessed Restoration" is not one of the least extraordinary. At all events, the retention of it as a sacred day is extraordinary, when we consider how long it is since the character of Charles and his court have been a by-word. But the custom was retained for the same reason that set it up—not to thank God, but to spite those who differed. The gusto of the gratitude was in proportion to the sufferings of the enemy. Cromwell thanked God for the head of Charles the First on a scaffold, and Charles the Second thanked God for the head of Cromwell on a gibbet. The defenders of the anniversary, if they spoke the truth, would have vindicated themselves on the plea that they did not thank God for Charles at all. To thank Him for Charles would have been to thank Him for Cleveland and Buckingham; for the pension from the French king, and all sorts of effronteries and enormities. Oh, no; the decorous men hated those. It was for no vice they fêted him. It was for the virtuous pleasure of galling their neighbors, and of doing honor to Mother Church herself, who condescended to be led back to her seat by the hand of the gay deceiver.

Now, Mother Church on that occasion was not the right, unpapal, unpuritanical unsophisticate Mother Church, old as no church at all, and ever young as advancement, but one of her spurious representatives; and society is awaking to the necessity of having no more such masqueraders, but seeing the beautiful, gentle, altogether Christian creature as she is, professing nothing that she does not believe, and believing nothing that can offend the wisest. Tillotson, Berkeley, Whichcote, have had sight of her. Charles the Second's chaplains knew no more of her than Dr. Philpotts.

THE SUN.

No mystery in creation need sadden us, as long as we believe nothing of the invisible world inferior to what the visible proclaims. Life and geniality predominate; death is brief; pain fugitive; beauty universal; order paramount and everlasting. What a shame, to know that the sun, the greatest visible object in our universe, combines equal gentleness with power, and does us nothing but good, and at the same time to dare to think worse of its Maker!

BON-MOT OF A COACHMAN.

Commendation beforehand is usually but a bad preface to a jest, or to anything else; yet I must say that I never heard anything more to the purpose than the reply made to a shabby fellow by the driver of an omnibus. SHABBY, on hailing the omnibus, had pathetically intimated that he had not more than a shilling, so that he could not pay the whole fare, which was eighteen pence. This representation *in formâ pauperis* the driver good-naturedly answered by desiring the gentleman to get in. The journey being ended, SHABBY, who had either been too loud in his pathos before the passengers, or too happy in the success of it, to think of getting change from them as he went (for it is manifest, from what followed, that he knew he had more than he pretended), was forced to develop from his purse a criminatory *half-crown!* This solid body of self-refutation, without pretending any surprise on his own part at the possession of it, and thus availing himself of an obvious opportunity, he hands to the coachman with a dry request for the difference. The coachman, still too good-natured to take any verbal notice of the pleasing apparition, but too wise not to do himself justice, returns twelve pence to SHABBY. SHABBY intimates his expectation of the sixpence.

COACHMAN. My fare, you know, sir, is eighteen pence.

SHABBY. Yes; but you said I was to ride for a shilling.

COACHMAN. I did; but you gave me to understand that you had no more in your pocket.

SHABBY. A bargain's a bargain.

COACHMAN. Well, then, sir, to tell you the truth, you must know that I am *the greatest liar on the road.*

SONG OF THE NIGHTINGALE.

The question respecting the mirth or melancholy of the nightingale, which of late years is supposed to have been settled in favor of the gayer side by some fine lines of Coleridge's, surely resolves itself into a simple matter of association of ideas. Chaucer calls the notes of the bird "merry"; but the word *merry*, in Chaucer's time, signified something alive and vigorous *after its kind;* as in the instance of "merry men" in the old ballads, and "merry England"; which did not mean a nation or set of men always laughing and enjoying themselves, but in good hearty condition—a state of manhood befitting men. This point is determined beyond a doubt by the same poet's application of the word to the organ, as the "merry organ"—meaning the *church*-organ, which, surely, however noble and organic, is not merry in the modern sense of the word.

The whole matter I conceive to be this. The notes of the nightingale, generally speaking, are not melancholy in themselves, but melancholy from association with night-time, and from the grave reflections which the hour naturally produces. They may be said to be melancholy also in the finer sense of the word (such as Milton uses in his "Penseroso"), inasmuch as they express the utmost intensity of vocal beauty and delight; for the last excessive feelings of delight are always grave. Levity does not do them honor enough, nor sufficiently acknowledge the appeal they make to that finiteness of our nature which they force unconsciously upon a sense of itself, and upon a secret feeling of our own capabilities of happiness compared with the brevity of it.

OVID.

Ovid was the son of a Roman knight, had an easy fortune, and (to use a modern phrase) was one of the gayest and most popular *men about town* in Rome for nearly thirty years; till, owing to some mysterious offense given to the court of Augustus, which forms one of the puzzles of biography, he was suddenly torn from house and home, without the least intimation, in the middle of the night, and sent to a remote and wintry place of exile on the banks of the Danube. Ovid was a good-natured man, tall and slender, with more affections than the levity of his poetical gallantry leads us to suppose. His gallantries are worth little, and have little effect; but his "Metamorphoses" are a store of beautiful Greek pictures, and tend to keep alive in grown people the feelings of their boyhood.

THE VOICE OF THE ROOK.

The Saxon word *rook* and the Latin word *raucus* (hoarse) appear to come from the same root;

though it is curious that neither Latins nor Italians have a name for the rook distinct from that of crow or raven, as the English have. The same sense, however, of the hoarseness of the bird's voice seems to have furnished the names of almost all the Corvican family—crow, rook, raven, daw, *corvus* and *cornix* (Latin), *korax* (Greek). When the rook is mentioned, nobody can help thinking of his voice. It is as much identified with him as bark with the old trees. But why do naturalists never mention the kindly *chuckle* of the young crows? particularly pleasant, good-humored, and infant-like; as different from the rough note of the elders as peel is from bark, or a baby's voice is from that of a man.

HOW LAWYERS GO TO HEAVEN.

There is a pleasant story of a lawyer, who, being refused entrance into heaven by St. Peter, contrived to throw his hat inside the door; and then, being permitted by the kind saint to go in and fetch it, took advantage of the latter's fixture as doorkeeper to refuse to come back again.

COLLINS THE POET.

In Mr. Pickering's edition of Collins there is an engraved likeness of the poet, the only one that has appeared. Nothing is said for its authenticity; it is only stated to be "from a drawing formerly in the possession of William Seward, Esq." ; but it possesses, I think, internal evidence of its truth, being clouded, in the midst of its beauty, with a look of pride and passion. There is also a thick-stuffed look in the cheeks and about the eyes, as if he had been overfed; no uncommon cause, however mean a one, of many a trouble in after-life.

The dreadful calamity which befell the poet has generally been attributed to pecuniary distresses occasioned by early negligence, or at least to habits of indolence and irresolution which grew upon him. His biographer, in this edition, says, with great appearance of justice, that the irresolution was always manifest; and he attributes the calamity to a weakness of mind that was early developed. But whence arose the weakness of mind? It is desirable, for the common interests of mankind, that biographers should trace character and conduct to their first sources; and it is little to say that a weakness was the consequence of a weakness. Collins's misfortune seems to have originated in the combined causes of delicacy of bodily organization, want of guidance on the part of relations, and perhaps in something of a tendency on their part to a similar malady. His father, a hatter, is described as being "a pompous man"; his sister pushed avarice and resentment to a pitch of the insane; the father died while his son was a boy, the mother not long afterward;

his uncle, Colonel Martin, though otherwise very kind, seems to have left him to his own guidance. The poet was so delicately organized, that in early life he expected blindness; and this ardent and sensitive young man, thus left to himself, conscious of great natural powers, which he thought he might draw upon at a future day, and possessing the natural voluptuousness of the poetical temperament, plunged into debt and pleasure beyond recovery, and thus, from a combination of predisposing circumstances, lost his wits. I think it discernible that he had his father's pride, though in better taste; and also that he partook of his sister's vehemence, though as generous as she was stingy. We learn from Sir Egerton Brydges, that, notwithstanding his delicacy of temperament, his shrieks were sometimes to be heard from the cloisters in Chichester to such an excess as to become unbearable. "Poor dear Collins!" we involuntarily exclaim with Dr. Johnson: how much we owe, pity, and love him! One can love any man that is generous; one pities Collins in proportion as he has taught us to love Pity herself; and I for one owe him some of the most delightful dreams of my childhood. Of my childhood, do I say? Of my manhood—of my *eternityhood*, I hope; for his dreams are fit to be realized in the next world.

"Thy form," says he, in his "Ode to Pity," speaking of the God of War—

> "Thy form from out thy sweet abode
> O'ertook him on his blasted road,
> And stopped his wheels, and looked his rage away!"

How did this passage, by the help of the pretty design by Mr. Kirk in Cooke's edition of the poets, affect me, and help to engage my heart for ever in the cause of humanity! An allegory may be thought a cold thing by the critics; but to a child it is often the best representation of the truth which he feels within him, and the man is so far fortunate who feels like the child. I used to fancy I saw Pity's house on the roadside—a better angel than those in Bunyan—and the sweet inmate issuing forth, on one of her dewy mornings, to look into the eyes of the God of War and turn him from his purpose.

If Collins had married and had a family, or been compelled to write not only for himself but others, it is probable that the morbidity of his temperament would have been spared its fatal consequences: the necessity of labor might have varied his thoughts, and sympathy turned his very weakness into strength. A good heart can hardly be conscious of belonging to many others, and not distribute itself, as it were, into their being, and multiply its endurance for their sake. But Collins might have had such an opinion of his disease as to think himself bound to remain single.

It does not appear that the greatest understandings, through whatever dangers they may pass from excess of thought, are liable to be finally borne down by it. They seize upon every help, and acquire the habit of conquest. But I suspect Collins to have been not only of a race overstocked with passion, but a spoiled child, habituated to the earliest indulgence of his feelings; and the infirmity may have become so strong for him as to render such a piece of self-denial at once the most painful and most reasonable of his actions. One retires with reverence before the possibility of such a trial of virtue; and can only end with hoping that the spirit which has given such delight to mankind is now itself delighted.

A FACT.

The powers of the printing-press are very extraordinary; yet the imaginations even of the dull can outstrip them. A woman, I have been told, absolutely went into a bookseller's shop, said she was going farther, and requested to have a Bible which should be "small in size, large in type, and printed by the time she came back." It was to a similar application that a bookseller replied: "I see what you want, madam; a pint-pot that will hold a quart." More things of this kind have been related, probably with truth; for there are as many strange truths of ignorance as of knowledge.

THE TWO CONQUERORS.

When Goethe says that in every human condition foes lie in wait for us, "vincible only by cheerfulness and equanimity," he does not mean that we can at all times be really cheerful, or at a moment's notice; but that the endeavor to look at the better side of things will produce the habit; and that this habit is the surest safeguard against the danger of sudden evils.

CLERICAL TITLES.

It is a pity that the clergy do not give up the solemn trifling of some of their titles. Their titular scales and gradations of merit become very ludicrous on inspection. Thus you may have a reverence for a curate of an apostolical life, supposing it possible to have it for a poor man; but you can have no *right* reverence. A bishop is the only man who is " Right Reverend." The curate can not even be " Venerable," however he may be venerated : it is the archdeacon that is Venerable. Again, a prebendary is not Most Reverend, though he is Very : the dean is the only man that is Most Reverend. There is a prevailing reverence in the prebendary : he is *valde reverendus ;* but the dean is filled and saturated and overflowing with venerability; he is superlatively reverend —*reverendissimus*. These distinctions often take place in the same man, in the course of a minute. An archdeacon for instance is dining, and has just swallowed his sixty-ninth mouthful; during which operation he was only Venerable. A messenger comes in, and tells him that he is a dean; upon which he spills the gravy for joy, and is Most Reverend.

HORACE WALPOLE AND PINKERTON.

Pinkerton was a man of an irritable and overweening mediocrity. His correspondence with Beattie, Percy, and others, is curious for little more than the lamentable evidence it affords of the willingness of grave men to repay the flatteries of a literary tyro, in a style which unquestionably did Mr. Pinkerton great harm in afterlife, and which is quite enough to account for the height of presumption to which it suffered his irritability to carry him. Those of Horace Walpole, who contributed to the mischief, are the best. Like all the letters of that celebrated person, whose genius was a victim to his rank, they are remarkable for their singular union of fine sense, foppery, and insincerity. He praises Mr. Pinkerton desperately at first; then gets tired of him, and mingles his praise with irony. Mr. Pinkerton finds out the irony, and complains of it; upon which the man of quality has the impudence to vow he is serious, and proceeds to hoax him the more.

One of Mr. Pinkerton's fantastic contrivances to supply his want of originality was a speculation for improving the English tongue by the addition of vowels to its final consonants. The number of final *s*'s in our language is certainly a fault. It is a pity we do not retain the Saxon plural termination in *en*, which we still have in the word *oxen*—as *housen* for *houses*, etc. But as changes for the worse grow out of circumstances, so must changes for the better; especially upon points on which the world can feel themselves but feebly interested. What would the Stock Exchange care for *consolso* instead of *consols?* or the poor for *breado*, if they could but get *bread?* or even a lover, who has naturally a propensity to soft words, for a *faira brida*, provided he has the lady? Yet upon improvements no wiser than these did Mr. Pinkerton and his correspondents busy themselves. One of them talks of *quieto nyto*, meaning a *quiet night;* and *honesta shepherda* and *shepherdeza!*

Pinkerton sometimes encouraged Walpole himself to get in a fantastic humor. Peter Pindar says:

"My cousin Pindar in his odes
Applauded horse-jockeys and gods."

Walpole expressed a serious opinion that a new Pindar might do likewise—that all the English games might be rendered poetical like those of the ancients; forgetting the differences of occasion, custom, religion, and a totally different state of society. A serious panegyric on a gentleman's horse might undoubtedly be well received by the owner, and the poet invited to dinner to hear a delicious conversation on bets and chances; but a ballad would do better than an ode. The latter would require translation into the vulgar tongue.

JEWS.

In our thoughts of old-clothesmen and despised shopkeepers, we are accustomed to forget that the Jews came from the East, and that they still partake in their blood of the vivacity of their Eastern origin. We forget that they have had their poets and philosophers, both gay and profound, and that the great Solomon was one of the most beautiful of amatory poets, of writers of epicurean elegance, and the delight of the whole Eastern world, who exalted him into a magician. There are plentiful evidences, indeed, of the vivacity of the Jewish character in the Bible. They were liable to very ferocious mistakes respecting their neighbors, but so have other nations been who have piqued themselves on their refinement; but we are always reading of their feasting, dancing and singing, and harping and rejoicing. Half of David's imagery is made up of allusions to these lively manners of his countrymen. But the Bible has been read to us with such solemn faces, and associated with such false and gloomy ideas, that the Jews of old become as unpleasant though less undignified a multitude in our imaginations as the modern. We see as little of the real domestic interior of the one as of the other, even though no people have been more abundantly described to us. The moment we think of them as people of the East, this impression is changed, and we do them justice. Moses himself, who, notwithstanding his share of the barbarism above mentioned, was a genuine philosopher and great man, and is entitled to our eternal gratitude as the proclaimer of the Sabbath, is rescued from the degrading familiarity into which the word Moses has been trampled, when we read of him in D'Herbelot as Moussa ben Amran; and even Solomon becomes another person as the Great Soliman or Soliman ben Daoud, who had the ring that commanded the genii, and sat with twelve thousand seats of gold on each side of him, for his sages and great men.

SMOLLETT.

Though Smollett sometimes vexes us with the malicious boy's-play of his heroes, and sometimes disgusts with his coarseness, he is still the Smollett whom now, as in one's boyhood, it is impossible not to heartily laugh with. He is an accomplished writer, and a masterly observer, and may be called the finest of caricaturists. His caricatures are always substantially true: it is only the complexional vehemence of his gusto that leads him to toss them up as he does, and tumble them on our plates. Then as to the objections against his morality, nobody will be hurt by it. The delicate and sentimental will look on the whole matter as a joke; the accessories of the characters will deter *them:* while readers of a coarser taste, for whom their friends might fear most, because they are most likely to be conversant with the scenes described, are, in our opinion, to be seriously benefited by the perusal; for it will show them that heroes of their description are expected to have virtues as well as faults, and that they seldom get anything by being positively disagreeable or bad. Our author's lovers, it must be owned, are not of the most sentimental or flattering description. One of their common modes of paying their court, even to those they best love and esteem, is by writing lampoons on other women! Smollett had a strong spice of pride and malice in him (greatly owing, we doubt not, to some scenes of unjust treatment he witnessed in early youth), which he imparts to his heroes; all of whom, probably, are caricatures of himself, as Fielding's brawny, good-natured, idle fellows are of *him.* There is no serious evil intention, however. It is all out of resentment of some evil, real or imaginary; or is made up of pure animal spirit and the love of venting a complexional sense of power. It is energy, humor, and movement, not particularly amiable, but clever, entertaining, and interesting, and without an atom of hypocrisy in it. No man will learn to be shabby by reading Smollett's writings.

CHEMISTRY.

We eat, drink, sleep, and are clothed in things chemical; the eye that looks at us contains them; the lip that smiles at the remark is colored by them; we shed tears (*horribile dictu!*) of soda-water. But we need not be humiliated. Roses and dew-drops contain the same particles as we; custom can not take away the precious mystery of the elements; the meanest compounds contain secrets as dignified as the most lofty. The soul remains in the midst of all, a wondrous magician, turning them to profit and beauty.

A good book about chemistry is as entertaining as a romance. Indeed, a great deal of romance, in every sense of the term, has always been mixed up with chemistry. This most useful of the sciences arose out of the vainest; at least *the art of making gold*, or the secret of the *philosopher's stone* (for chemistry originally meant nothing more), has hitherto had nothing to show for itself but quackery and delusion. What discoveries the human mind may arrive at, it is impossible to say. I am not for putting bounds to its possibilities, or saying that no Columbuses are to arise in the intellectual world, who shall as far surpass the other as the universe does our hemisphere. But meanwhile chemistry supplied us with food for romances before it took to regulating that of the stomach, or assisting us in the conquest of the world material. We owe to it the classification and familiar intimacy of the Platonical world of spirits, the Alchemists of Chaucer and Ben Jonson, partly even of the "Rape of the Lock." Paracelsus's Dæmon of the stomach was the first that brought the spiritual and medical world into contact: in other words, we owe to that extraordinary person, who was an instance of the freaks played by a great understanding when it is destitute of moral sensibility, the first application of chemical knowledge to medicine. The amiable and delightful Cullen, in whom an extreme humanity became a profound wisdom (and the world are still to be indebted to him in morals as well as physics), was the first who enlarged the science into the universal thing which it is now. This was not a hundred years ago. To what a size has it not grown since, like the vapory giant let out of the casket!

PETTY CONVENIENCES AND COMFORTS.

The locks and keys, and articles on a par with them, in Tuscany, are, perhaps, the same now that they were in the days of Lorenzo de' Medici. The more cheerful a nation is in ordinary, or the happier its climate, the less it cares for those petty conveniences which irritable people keep about them, as a set-off to their want of happiness in the lump. A Roman or a Tuscan will be glad enough to make use of an English razor when he gets it; but the point is, that he can do better without it than the Englishman. We have sometimes seen in the face of an Italian, when English penknives and other perfections of manufacture have been shown him, an expression, mixed with his wonder, of something like paternal pity, as if the excess of the thing was childish. It seemed to say: " Ah, you can make those sort of things, and we can do without them. Can you make such knicknacks as Benvenuto Cellini did—carcanets and caskets, full of exquisite sculpture, and worth their weight in jewels?"

And there is reason in this. It is convenient to have the most exquisite penknives; but it is a greater blessing to be able to do without them. No reasonable man would stop the progress of manufacture, for a good will come of it beyond what was contemplated. But it is not to be denied, meanwhile, that the more petty conveniences we abound in, the more we become the slaves of them, and the more impatient at wanting them where they are not. Not having the end, we keep about us what we take for the means. Cultivators of better tempers or happier soils get at the end by shorter cuts. The only real good of the excessive attention we pay to the conveniences of life is, that the diffusion of knowledge and the desire of advancement proceed in company with it; and that happier nations may ultimately become still happier by our discoveries, and improve us, in their turn, by those of their livelier nature.

TEARS.

Sympathizing and selfish people are alike given to tears, if the latter are selfish on the side of personal indulgence. The selfish get their senses into a state to be moved by any kind of excitement that stimulates their languor, and take a wonderful degree of pity on themselves; for such is the secret of their pretended pity for others. You may always know it by the fine things they say of their own sufferings on the occasion. Sensitive people, on the other hand, of a more generous sort, though they can not always restrain their tears, are accustomed to do so, partly out of shame at being taken for the others, partly because they can less afford the emotion. The sensitive selfish have the advantage in point of natural strength, being often as fat, jolly people as any, with a trick of longevity. George IV., with all his tears, and the wear and tear of his dinners to boot, lasted to a reasonable old age. If he had been shrewder, and taken more care of himself, he might have lived to a hundred. But it must be allowed that he would then have been still more selfish than he was; for these luxurious weepers are at least generous in imagination. They include a notion of other people somehow, and are more convertible into good people when young. The most selfish person we ever met with was upward of a hundred, and had the glorious reputation of not being movable by anything or anybody. He lasted as a statue might last in a public square, which would see the whole side of it burn with moveless eyes and bowels of granite."

DR. ALDRICH.

Aldrich, Dean of Christ Church, built some well-known and admired structures at Oxford; was a musician as well as architect; wrote the famous "Smoking Catch" (being accomplished in the smoking art also); was the author of "Hark! the Bonny Christ Church Bells," a composition of great sprightliness and originality; and has the reputation of being an elegant Latin poet. His Latin verses are to be met with in the "Musæ Anglicanæ"; but we do not remember them, unless the following hexameters be among the number:

"Si bene quid memini, causæ sunt quinque bibendi:
Hospitis adventus, præsens sitis, atque futura,
Aut vini bonitas, aut quælibet altera causa."

Which has been thus translated, perhaps by the author, for the version is on a par with the original:

"If on my theme I rightly think,
There are five reasons why men drink:
Good wine, a friend, or being dry,
Or lest we should be by-and-by,
Or any other reason why."

LORD MARCHMONT'S RECEIPT FOR LONGEVITY.

Lord Marchmont, the friend of Pope, lived to the age of eighty-six, preserving his strength and faculties to the last. He rode out only five days before he died. Sir John Sinclair, who knew him, wished to ascertain the system he pursued, and received for answer that his lordship always lived as other people did, but that he had laid down when young one maxim, to which he rigidly adhered, and to which he attributed much of his good health, namely:

Now, what do you think this maxim was? Never to exceed in his eating? No. Never to lie late in bed? No. Never to neglect exercise?

Never to take much physic? Never to be rakish, to be litigious, to be ill-tempered, to give way to passion? No, none of these. It was

Never to mix his wines.

What luxurious philosophies some people have! My Lord Marchmont was resolved to be a long-lived, virtuous, venerable man; and therefore he laid it down as a maxim, Never to mix his wines. To get one glass of wine, in their extreme weakness, is what some human beings, bent double with age, toil, and rheumatism, can seldom hope for; while another of the race, having nothing to bend him and nothing to do, shall become a glorious example of the beauty of this apostolical maxim — "Never to mix your wines." Lord Marchmont did accordingly for many years generously restrict himself to the use of claret; but his physicians having forbidden him to take that wine on account of its acidity, he resolved, with equal self-denial, to "confine himself to Burgundy"; and accordingly, with a perseverance that can not be sufficiently commended, he "took a bottle of it every day for fifteen years."

The noble lord was a good man, however, and his "neat, as imported," is not to be grudged him. All we have to lament is, that thousands, as good as he, have not an atom either of his pleasure or his leisure.

THE AMERICAN REVOLUTION.

There is something in the history of the American Revolution extremely dry and unattractive. This is owing partly perhaps to the moneyed origin of it, partly to the want of personal anecdotes, to the absence of those interesting local and historical associations which abound in older states, and to the character of Washington; who, however admirable a person, and fitted as if by Providence to the task which he effected, was himself, personally, of a dry and unattractive nature, an impersonation of integrity and straightforwardness, exhibiting none of the social or romantic qualities which interest us in other great men. For similar reasons, the American Indians are the least interesting of savages. Their main object has been to exhibit themselves in an apathetic or stoical character, and they have suffered in human sympathy accordingly.

DISCOVERERS OF AMERICA.

It is painful to reflect on the calamitous circumstances under which these high-minded adventurers were accustomed to terminate their careers, however brilliant their successes by the way. They got riches and territory for others, and generally died in poverty, often of wounds and disease, sometimes by the hands of the executioner. Pinzon, who first crossed the equinoctial line in the New Hemisphere and discovered Brazil, got nothing by his voyage of discovery but heavy losses. Nicuesa disappeared, and was supposed to have perished at sea. Valdivia was killed and eaten by cannibals. Ponce de Leon, who thought to discover the fountain of youth, died of a wound exasperated by mortified pride and disappointment. The lofty and romantic Don Alonzo de Ojeda died so poor that he did not leave money enough to provide for his interment; and so broken in spirit that with his last breath he entreated that his body might be buried in the monastery of San Francisco, just at the portal, in humble expiation of his past pride, that "every one who entered might tread upon his grave." And Vasco Nuñez de Balboa, one of the best of the old brotherhood, perished on the scaffold, a victim, like Columbus, to envy. It is to be recollected, however, that such men accomplish the first object of their amibition—renown; and that life, and not death, is the main thing by which we are to judge of their happiness.

WONDER NEVER CEASES.

It might be thought that the progress of science would destroy the pleasure arising from the perusal of works of fiction, by showing us the mechanical causes of phenomena, and so leading us to conclude that the utmost wonders we could imagine might with equal reason be referred to similar causes. In other words, no wonder is greater than any other wonder; and, if once explained, it ceases to be a wonder. "Wonder," it has been said, "is the effect of novelty upon ignorance." Perhaps it would have been said better, that wonder is the effect of want of familiarity upon ignorance: for there are many things that excite our wonder, though far from new to us or to our reflections; such as life and death, the phenomena of the planets, etc. But to say nothing of the inexhaustible stock of novelties, wonders could never cease in anything till we knew their first as well as their final causes. We must understand how it is that substance, and motion, and thought exist, before we can cease to admire them: the very power of writing a fairy tale is as great a wonder as anything it relates. And thus, while we think to frighten away the charms of fable and poetry with the sound of our shuttles and steam-engines, they only return the more near to us, settle smiling on the very machinery, and (to say nothing of other sympathies) demand admiration on the very same grounds.

DALY, THE DUBLIN MANAGER.

Daly, patentee of the Dublin Theatre, was one of those iron-hearted and brazen-faced blackguards, who, in an age when knowledge is on the increase, are not so likely to be taken for clever fellows as they used to be; being in fact no other than scoundrels in search of a sensation, and willing to gratify it, like wild beasts, at the risk of any price to the sufferer. Such fellows do not abound with courage: they merely have one of an honorable man's drawbacks upon ferocity. To talk of their other gallantry would be equally preposterous. Even of animal impulse they know no more than others. They only know no restraint. Give a man good health, and take from him all reflection and every spark of love, and you have the human wild beast called Daly. His best excuse was his squint. There was some smack of salvation in that, for it looks as if he resented it.

"Richard Daly, Esq., patentee of the Dublin Theatre" (says Boaden's "Life of Mrs. Jordan"), "was born in the county of Galway, and educated at Trinity College. As a preparation for the course he intended to run through life, he had fought sixteen duels in two years, three with the small-sword and thirteen with pistols; and he, I suppose, imagined, like Macbeth, with equal confidence and more truth, that he bore a 'charmed life'; for he had gone through the said sixteen trials of his nerve without a single wound or scratch of much consequence. He, therefore, used to provoke such meetings on any usual and even uncertain grounds, and entered the field in pea-green, embroidered and ruffled and curled, as if he had been to hold up a very different ball, and gallantly presented his full front, conspicuously finished with an elegant brooch, quite regardless how soon the labors of the toilet 'might soil their honors in the dust.' Daly, in person, was remarkably handsome, and his features would have been agreeable but for an inveterate and most distressing squint, the consciousness of which might keep his courage eternally upon the lookout for provocation, and not seldom, from surprise alone, afford him an opportunity for this his favorite diversion. Like Wilkes, he must have been a very unwelcome adversary to meet with the sword, because the eye told the opposite party none of his intentions. Mr. Daly's gallantry was equal at least to his courage, and the latter was often necessary to defend him in the unbridled indulgence that through life he permitted to the former. He was said to be the general *lover* in his theatrical company; and, I presume, the resistance of the fair to a *manager* may be somewhat modified by the danger of offending one who has the power to appoint them to parts, either striking or otherwise, and who must not be irritated if he can not be obliged. It has been said, too, that any of his subjects risked a great deal by an escape from either his love or his tyranny; for he would put his *bond* in force upon the refractory, and condemn to a hopeless imprisonment those who, from virtue or disgust, had determined to disappoint him."

LIGHT AND COLORS.

Light is, perhaps, the most wonderful of all visible things; that is to say, it has the least analogy to other bodies, and is the least subject to secondary explanations. No object of sight equals it in tenuity, in velocity, in beauty, in remoteness of origin, and closeness of approach. It has "no respect of persons." Its beneficence is most impartial. It shines equally on the jewels of an Eastern prince and on the dust in the corner of a warehouse. Its delicacy, its power, its utility, its universality, its lovely essence, visible and yet intangible, make up something godlike to our imaginations; and, though we acknowledge divinities more divine, we feel that ignorant as well as wise fault may be found with those who have made it an object of worship.

One of the most curious things with regard to light is, that it is a body, by means of which we become sensible of the existence of other bodies. It is a substance; it exists as much in the space between our eyes and the object it makes known to us as it does in any other instance; and yet we are made sensible of that object by means of the very substance intervening. When our inquiries are stopped by perplexities of this kind, no wonder that some awe-stricken philosophers have thought further inquiry forbidden; and that others have concluded, with Berkeley, that there is no such thing as substance but in idea, and that the phenomena of creation exist but by the will of the Great Mind, which permits certain apparent causes and solutions to take place, and to act in a uniform manner. Milton doubts whether he ought to say what he felt concerning light:

> "Hail, holy Light, offspring of Heaven first-born,
> Or of the eternal coeternal beam,
> May I express thee unblamed? since God is light,
> And never but in an unapproached light
> Dwelt from eternity, dwelt there in thee,
> Bright effluence of bright essence increate."

And then he makes that pathetic complaint, during which we imagine him sitting with his blind eyes in the sun, feeling its warmth upon their lids, while he could see nothing:

> ". . . Thee I revisit safe,
> And feel thy sovran vital lamp; but thou
> Revisit'st not these eyes, that roll in vain
> To find thy piercing ray, and find no dawn."

As color is imparted solely by the different rays of light with which they are acted upon, the sun *literally* paints the flowers. The hues of the pink and rose literally come, every day, *direct from heaven.*

VERSIONS OF ANCIENT LYRICS.

The more we consider Anacreon and the ancient lyrics, the more probable it seems that some degree of paraphrase is necessary, to assimilate them in effect to the original. We are to recollect that the ancient odes were written to be sung to music; that the poet himself was the first performer; and that the idea of words and music was probably never divided in the mind of the reader. The spirit of enjoyment is a spirit of continuousness. We may suppose what we like of Greek simplicity and brevity, especially in their epigrams or inscriptions, the shortness of which was most likely prescribed, in the first instance, by the nature of the places on which they were written; but we may be pretty certain that the shortest of Anacreon's songs was made three, or four, or five times as long as it appears to us, by the music with which it was accompanied. Take a song of Metastasio's, as set by Arne or Mozart, and we shall find the duration of it a very different thing in the study and the theatre. The only true way, therefore, of translating an ode of Anacreon, is to sympathize as much as possible with his animal spirits, and then to let the words flow as freely as they will, with as musical and dancing a melody as possible, so as to make the flow and continuity of the verse as great a substitute as possible for the accompaniment of the lyre.

The only versions of Anacreon in the English language that are really worth anything, are those of Cowley; and these are as paraphrastic as they are joyous.

CATHARINE II. OF RUSSIA.

As long as she had everything her own way, Catharine could be a very pleasant, vain, debauched, fat-growing, all-tolerant mistress, interchanging little homages with the philosophers; but as soon as philosophy threatened to regard the human race as of more consequence than one woman, adieu to flattery and to France. The French then were only worthy of being "drubbed."

Catharine was a clever *German*, with a great deal of will, among a nation of barbarians. This is the clew to her ascendancy. In a more southern country she would probably have been little thought of, in comparison with what she was reputed as the "mother" of her great clownish family of Russians.

Note.—That the arbitrary have always had a tendency to grow fat, for the same reason that inclines them to be furious. The same people who can deny others everything are famous for refusing themselves nothing.

PETRARCH AND LAURA.

There is plenty of evidence in her lover's poetry to show that Laura portioned out the shade and sunshine of her countenance in a manner that had the instinctive effect of artifice, though we do not believe there was any intention to practice it. And this is a reasonable conclusion, warranted by the experience of the world. It is not necessary to suppose Laura a perfect character, in order to excite the love of so imaginative a heart as Petrarch's. A good half or two thirds of the love may have been assignable to the imagination. Part of it was avowedly attributable to the extraordinary fidelity with which she kept her marriage vow to a disagreeable husband, in a city so licentious as Avignon, and, therefore, partook of that not very complimentary astonishment and that willingness to be at an unusual disadvantage, which make chastity cut so remarkable a figure amid the rakeries of Beaumont and Fletcher. Furthermore, Laura may not have understood the etherealities of Petrarch. It is possible that less homage might have had a greater effect upon her; and it is highly probable (as Petrarch, though he speaks well of her natural talents, says she had not been well educated) that she had that instinctive misgiving of the fine qualities attributed to her, which is produced even in the vainest of women by flights to which they are unaccustomed. It makes them resent their incompetency upon the lover who thus strangely reminds them of it. Most women, however, would naturally be unwilling to lose such an admirer, especially as they found the admiration of him extend in the world; and Laura is described by her lover as manifestly affected by it. Upon the whole, I should guess her to have been a very beautiful, well-meaning woman, far from insensible to public homage of any sort (she was a splendid dresser, for instance), and neither so wise nor so foolish as to make her seriously responsible for any little coquetries she practiced, or wanting in sufficient address to practice them well. Her history is a lofty comment upon the line in "The Beggars' Opera"—

"By keeping men off, you keep them on."

As to the sonnets with which this great man immortalized his love, and which are full of the most wonderful beauties, small and great (the versification being surprisingly various and charming, and the conceits of which they have been accused being for the most part as natural and delightful as anything in them, from a propensity which a real lover has to associate his mistress with everything he sees), justice has been done to their gentler beauties, but not, I think, to their intensity and passion. Romeo should have written a criticism on Petrarch's sonnets. He would have done justice both to their "conceits" and their fervor. I think it is Ugo Foscolo who remarks that Petrarch has given evidence of passion felt in solitude, amounting even to the terrible. His temperament partook of that morbid cast which makes people haunted by their ideas, and which, in men of genius, subjects them sometimes to a kind of delirium of feeling, without destroying the truth of their perceptions. Petrarch more than once represents himself in these sonnets as struggling with a propensity to suicide; nor do we know anything more affecting in the record of a man's struggles with unhappiness than the one containing a prayer of humiliation to God on account of his passion, beginning

> "Padre del ciel, dopo i perduti giorni"—
> (Father of heaven, after the lost days).

The commentators tell us that it was written on a Good Friday, exactly eleven years from the commencement of his love.

MORAL AND PERSONAL COURAGE.

In all moral courage there is a degree of personal; personal is sometimes totally deficient in moral. The reason is, that moral courage is a result of the intellectual perceptions and of conscience; whereas a man totally deficient in those may have nerves or gall enough to face any danger which his body feels itself competent to oppose. When the physically courageous man comes into the region of mind and speculation, or when the question is purely one of right or wrong, he is apt to feel himself in the condition of the sailor who confessed that he was afraid of ghosts, because he "did not understand their tackle." When moral courage feels that it is in the right, there is no personal daring of which it is incapable.

TIGHT-LACING.

It is a frequent matter of astonishment why females should persist in tight-lacing when so much is said against it, and how it happens that they should take what is really a deformity for something handsome. The first part of this mystery is answered by the second: they think the waist produced by tight-lacing a beauty; and the reason why they think so is, that they know a small waist to be the object of admiration, and they feel that they can never persuade you it is small without forcing the smallness upon your eyes, and thus forcing you to acknowledge it. On the contrary, the spectator feels that, if the waist were really small, so much pains would not be taken to convince him of it. But this the poor creatures will not consider. Every one thinks that there will be an exception in her favor. *Other* women, she allows, make themselves ridiculous, and attempt to impose upon us; with herself the case is different: everybody must see that *her* waist is really small. Therefore she goes lacing and lacing on, till she becomes like a wasp; and everybody who follows her in the street laughs at her.

Some of these waists are of such frightful tenuity as to strike the least thinking observer with their ugliness. The other day there was a young lady walking before me in the street, whose waist literally seemed no thicker than a large arm. The poor girl had marked herself for death. Some of the most vital parts of her body must have been fairly lapped over one another, or squeezed into a mass. My first sensation, on seeing this phenomenon, was horror at the monstrosity; the second was vexation with the poor silly girl; the third was pity. The ground of the stupid custom is sympathy, however mistaken. The poor simpletons wish for our admiration, and do not know how hard they try to gain our contempt. We ought to be the less provoked, because in all these yearnings after social approbation there is the germ of a great preferment for the community; since the same people who now make themselves so ridiculous, and get so much death and disease, by pursuing false means of obtaining our good opinion, would, in a wiser state of society, be led as vehemently to adopt the true. Instead of going about half stifled with bad vitals and ready-made coroner's inquests, the poor creatures would then be anxious to show us that they were natural healthy females, fit to be wives and mothers. At present, if they can be mothers at all, it is frightful to think what miseries they may inflict on their offspring.

GRAVITY AND INDUSTRY OF DANCERS.

One of Addison's happy papers in the "Spectator" (and how numerous they are!) contains an account of a mysterious personage who, lodging at the same house as his observer and making a great noise one day over his head, was watched by some of his fellow lodgers through the key-hole. They observed him look gravely on a book, and then twirl round upon one leg. He looked gravely again, and put forth his leg in a different manner. A third time he fell to studying profoundly, and then, darting off with vivacity, took a career round the room. The conclusion was, if I remember, that he was some mad gentleman. The peepers, however, ventured in, and upon inquiry found that he was a dancing-master. The Spectator, who had joined them, concluded by requesting that the gentleman would be pleased in future to addict himself with less vehemence to his studies, since they had cost him that morning the loss of several trains of thought, besides breaking a couple of tobacco-pipes.

They who have seen the grave faces and lively legs of some of the opera-dancers, can easily understand the profundities of this master of their art; nor will they fall into the mistake of young people in supposing that a dancer has nothing to do but to be lively and enjoy himself. M. Blasis, the author of a work on the art, says that the dancer must be always practicing, otherwise he is in danger of losing what he has acquired. Some muscle will get out of practice, some shiver of the left leg be short of perfection. Furthermore, he must follow neither "simple unpracticed theorists," nor the "imaginary schemes of innovating speculators." He must also be temperate and sober; nay, must "partially renounce every pleasure but that which Terpsichore affords"; must not think of horsemanship, fencing, or running; must study the antique, drawing, and music, but particularly his own limbs; and, if he aspire to the composition of ballets, must have a profound knowledge of the drama and of human nature. See now, you who reflect but little, how much it takes to bring a man to a right state of *pirouette;* what world of accomplishment there is in that little toe, which seems pointed at nothing; and what a right the possessor of it has to the grave face which has so often puzzled conjecture. He seems to be merely holding the tip of a lady's finger; but who is to know what is passing through his mind?

"Use your endeavors," saith Blasis, "to twirl delicately round on the points of your toes." Here we feel in a state of anxiety, with a world of labor before us. In another sentence, one hardly knows in what sense we are to take his words—whether as an encouragement to tranquillity of mind, or an injunction to acquire lissomness in the body. "Make yourself easy," quoth he, "about your hips."

ADVERTISEMENTS.

Advertisements are sometimes very amusing. They give insights into the manners of the times no less interesting than authentic. Suppose the ancients had possessed a press, and that a volume of a Roman "Post" or "Chronicle" had been dug up at Herculaneum, with what curiosity should we not contemplate the millinery of the Roman ladies, or, "Wanted, a Gladiator to fight the last new lion"; or, "Next Ides of November will be published the new poem of Quintus Horatius Flaccus"; or a long account of a court-day of Nero or Antoninus! The best editions of the "Tatler" and "Spectator" have very properly retained a selection of the Advertisements.

SPORTSMEN AND CUSTOM.

There are unquestionably many amiable men among sportsmen, who, as the phrase is, would not "hurt a fly"—that is to say, on a window. At the end of a string, the case is altered. So marvelous are the effects of custom and education. Consoling thought, nevertheless! for if custom and education have been so marvelous in reconciling intelligent men to absurdities, and humane men to cruelty, what will they not effect when they shall be on the side of justice? when reason, humanity, and enjoyment shall become the three new graces of the civilized world? It has been said that absurdities are necessary to man; but nobody thinks so who is not their victim. With occupation, leisure, and healthy amusement, all the world would be satisfied.

BEARS AND THEIR HUNTERS.

It is natural in bear-hunters, who have witnessed the creature's ravages, and felt the peril of his approach, to call him a ferocious animal, and gift him at times with other epithets of objection; but we who sit in our closets, far removed from the danger, may be allowed to vindicate the character of the bear, and to think that Bruin, who is only laboring in his vocation, and is not more ferocious than hunger and necessity make him, might, with at least equal reason, have advanced some objections against his invader. He might have said, if he possessed a little Æsopean knowledge of mankind: "Here, now, is a fellow coming to kill me for getting my dinner, who eats slaughtered sheep and lobsters boiled alive; who, with the word 'ferocity' in his mouth, puts a ball into my poor head, just as the highwayman vindicates himself by abusing the man he shoots; and who then writes an account of his humane achievement with a quill plucked from the body of a bleeding and screaming goose."

Or, knowing nothing of mankind, he might say: "Here comes that horrid strange animal to murder us, who sometimes has one sort of head and sometimes another (hat and cap), and who carries another terrible animal in his paw — a kind of stiff snake — which sends out thunder and lightning; and so he points his snake at us, and in an instant we are filled with burning wounds, and die in agonies of horror and desperation."

There is much resemblance to humanity in the bear. I would not make invidious comparisons; but travelers as well as poets have given us beautiful accounts of the maternal affections of the bear. And furthermore, the animal resembles many respectable gentlemen whom we could name. When he wishes to attack anybody, he rises on his hind legs, as men do in the House of Commons; he dances, as aldermen do, with great solemnity and weight; and his general appearance, when you see him walking about the streets with his keeper, is surely like that of many a gentleman in a great-coat, whose enormity of appetite and the recklessness with which he indulges in it entitle him to have a keeper also.

SELF-STULTIFICATION.

The highest, most deliberate, peremptory, and solemn instance perhaps on record of this species of absurdity, is the dismissal of his court-fool, Archibald Armstrong, by Charles I. in council. Archy, as he was called, had given mortal offense to Laud by ridiculing his attempts at church-domination. It is related of him that he once said, by way of grace before dinner, "Great praise to the King, and Little Laud to the devil." But the last feather that broke the back of the Archbishop's patience was Archibald saying to him, on the failure of his liturgy in Scotland, "Who's fool now?" Laud complained to Charles; Charles summoned his council to take cognizance of the dreadful matter; and accordingly, at "Whitehall, on the eleventh of March, one thousand six hundred and thirty-seven, present the King's Most Excellent Majesty, the Lord Archbishop of Canterbury, the Lord Keeper, Lord Treasurer, Lord Privy Seal," and fourteen other great personages, Archibald Armstrong, "the King's fool," for certain scandalous words, of a "high nature," and "proved to be uttered by him by two witnesses," was sentenced to have "his coat pulled over his ears," and discharged from his Majesty's service.

What was this but saying that the fool was a fool no longer? "Write me down an ass," says Dogberry in the comedy. Write down that Archy is no fool, says King Charles in council; he has called the Archbishop one; and therefore we are all agreed, the Archbishop included, that the man has proved himself to be entitled no longer to the appellation.

COWSLIPS.

A country-girl the other day expressed her astonishment that ladies could see anything to admire in "cowslips." Now, here was an instance of the familiarity that breeds contempt. Cowslips are among the most elegant of the spring flowers. They look, with those pretty sleeves of theirs, like ladies themselves in their morning dresses. But the country-girl had been accustomed to see whole fields of them, and to associate them with wet and mire, and Farmer Higgins.

Shakespeare mentions cowslips seven times, primroses just as often, and violets fourteen. He says nothing of anemones or hyacinths. I gather this from Mrs. Clarke's "Concordance," which, besides being admirably what it professes to be, suggests curious speculations as to the greater or less likings of Shakespeare, his habitual associations of ideas, etc.; and it might be made subservient to interesting inquiries on those subjects.

APRIL FOOLS.

An anniversary of this kind, in which stultification is the order of the day, appears to take place about the same time of the year all over the civilized world. Yet it would look more like a custom originating in some one particular country than most of those which are thought to have had such commencements; for it is as difficult not to imagine ordinary holidays and superstitions the natural growth of every human community as it would be to suppose that all the world, at one particular season, agreed to make fools of one another without knowing it.

There are solemn people whose dignity can not bear to be disturbed, let the season be never so full of gayety. It is such a fragile and empty pretension, they are afraid that the least touch will knock it to pieces. Not so with the wiser. They rejoice in every good which Nature has bestowed on them, mirth included; and are only balked by the presence of the incompetent. The celebrated Dr. Clarke was once amusing himself at some merry pastime with some youths of his college, when he suddenly left off at the sight of one of the fellows. "Hush, boys," said he, "we must be quiet. Here's a fool coming."

I must tell you a story of a friend of mine, which I take to be a crowning specimen of April-fool making.

Down comes this father of a family one April day to breakfast, with a face looking at once amused and confounded, as if something had happened to him both pleasant and mortifying. The mother of the family asks the reason, and all his children's eyes are turned on him. He looked at first as if he did not like to speak; but on being pressed assumed an aspect of bold acknowledgment, and said, "Well, my dear, you know I am not particular on April days, but certainly I did not think that Harriet (one of the servants) would have gone so far as this."

"What is it?"

"Why, she has made an April fool of *me!!* I was coming down the stairs, when she requested me to have a care of a broom that was lying at the bottom of it. There was no broom, and she ran away laughing."

"Well," cries the lady, "of all the bold girls I ever met in my life, that Harriet has the greatest effrontery."

The children all joined in the astonishment. They never heard of such a thing. It was wonderful, shameful, etc.; but they could not help laughing, and the roar became universal.

"My dear," said Harry gravely, "and you, all of you merry young ladies and gentlemen, I have the pleasure of informing you, all round, and at one fell swoop, that you are a parcel of April fools."

PRIVATE WAR.

In the times when duels were fought with swords, the Dutch had a pretty custom (perhaps have it still in sequestered places, where virtue survives) in which two rustical parties, whenever they happened to have an argument over their beer, and couldn't otherwise settle it, took out the knives with which they had been cutting their bread and cheese, and went to it like gentlemen. It was called *snick-and-snee*, which is understood to mean *catch and cut*, the parties catching hold of one another by the collar or waistcoat, and thus conveniently *sneeing* or cutting away, as butchers might do at a carcass. A similar custom is related of the Highlanders, who, whenever they sat down together to dinner, were so prepared for it that in case of accidents, that is to say, of arguments, they stuck their dirks into the board beside their trenchers, so as to have their reasons ready at hand. If a man said, "You grow hot and ridiculous," out came the cold steel to disprove his words; and the question was settled upon the most logical military principles.

Now, if private and public virtue are identical, as moralists insist they are, in contradiction to the casuists of expediency, there is no reason why the disputes of individuals should not be settled like those of nations, in the good old Dutch and Highland manner. But, at the same time, as moralists and casuists alike agree in thinking that the more the system of war can be humanized the better, I can't but think that an obvious mode presents itself of showing the resort to bloodshed in its best and most reasonable colors—a light at once conclusive and considerate, humane yet valiant, elegant in the accessories, yet as *no-nonsense* and *John-Bull* like as the perfection of reason can desire. War, observe, is a very filthy as well as melancholy thing. There can be no doubt of that. And, therefore, on the *no-nonsense* principle, the fact is not to be disguised. People, it is true, do disguise it; writers of dispatches disguise it; even Wellington says little or nothing about it, which I have always thought the only blot on the character and candor of that great man. But I am sure that, on reflection, and considering how *un-English-like* such insincerity is, the Duke would give up the concealment after his usual manly fashion.

My plan is this: that whenever two gentlemen, alive to the merits and necessities of war, should happen to have a dispute over their wine, they should immediately put on two laced hats, call in a band of music from the streets, and after hearing a little of it, and marching up and down the room with an air of dignified propriety, fall to it with their fists, and see which can give the other the most logical bloody nose. The sight of blood adding to the valor of the combatants, the noses of course would get worse and worse, and the blows heavier and heavier, till both of the warriors reasonably became "sights," and one of the two at last fell insensible—that being an evil necessary to the termination of the argument. Meantime, they would groan considerably, and complain in a very touching manner of the kicks and cuffs they received on the tenderest parts of their bodies (to show that there was "no nonsense"); a great dust would be struck up from the carpet; pools of blood would properly overflow it (always to show that there was "no nonsense"); and then, when the fight was over, and the band of music had played again, and the shrieks in the drawing-room and kitchen had subsided into those tears and sobs which are the final evidences of a state of logical conviction, the conqueror (if he was able), or his friends at all events, would clear their throats in the most dignified manner, strike up a hymn, and thank the Author of their respective vitalities that the defeated party had been beaten to a jelly, to the special satisfaction of the beater, and the eternal honor and glory of the Author of the Universe.

N. B.—You must be cautious how you doubt whether the Author of the Universe takes any particular notice of the bloody noses, or whether he does not rather leave them to work out some different third purpose by themselves; because, in that case, you might be charged with wanting a due sense of his dignity. On the other hand, you must not at all imagine that he approves the bloody noses in the abstract as well as concrete; because, in that case, you would be charged with doubting his virtue. And, again, you are not to fancy that Heaven wishes to put an end to the bloody noses altogether; for that would be quite opposed to the principle of "no nonsense."

Your business is to preach love to your neighbor, to kick him to bits, and to thank God for the contradiction.

BEAUMARCHAIS.

Beaumarchais, author of the celebrated comedy of "Figaro," an abridgment of which has been rendered more celebrated by the music of Mozart, made a large fortune by supplying the American republicans with arms and ammunition, and lost it by speculations in salt and printing. His comedy is one of those productions which are accounted dangerous, from developing the spirit of intrigue and gallantry with more gayety than objection; and they would be more undeniably so, if the good humor and self-examination to which they excite did not suggest a spirit of charity and inquiry beyond themselves.

MOZART.

Mozart is wonderful for the endless variety and undeviating grace of his invention. Yet his wife said of him, that he was a still better dancer than musician! In a soul so full of harmony, kindness toward others was to be looked for; and it was found. When a child, he would go about asking people "whether they loved him." When he was a great musician, a man in distress accosted him one day in the street; and, as the composer had no money to give him, he bade him wait a little, while he went into a coffee-house, where he wrote a beautiful minuet extempore, and, sending the poor man with it to the music-seller's, made him a present of the proceeds. This is the way that great musicians are made. Their sensibility is their genius.

VIOLET—WITH A DIFFERENCE.

"Violet" is thought a suitable name for the sweetest heroines of romance, on account of its association with the flower; yet add but a letter to it and that not a harsh one, and it becomes the most unfeminine of characteristics—Violent.

VERBAL MISTAKES OF FOREIGNERS.

The Abbé Georgel, having to send a dinner-invitation to Hume from Prince Louis de Rohan, took the opportunity of impressing the historian with his knowledge of the English language in the following terms :

"*M. l'Abbé Georgel fait un million de complimens à M. Hume. He makes great account of his vorks, admires her wit, and loves her person.*"

If ever Hume shook his fat sides with laughter, it must have been at the English of M. l'Abbé Georgel. There is an old joke on the coast of France about an English lady, who, in putting up at an inn, raised a great confusion in the minds of the attendants by showing herself very particular about her two " sailors " (*matelots*); when all that she meant to impress was her nicety respecting two " mattresses " (*matelas*). The Italians have similar jokes about Englishmen declining to have any more at dinner, because they have eaten " ships " (the term for which, *bastimento*, they mistake for *abbastanza*, enough); upon which another declines too, on the ground that he had eaten " the anchor " (pronouncing *áncora* instead of *ancòra*, also). I remember an English lady in Italy, who became accomplished in the language; but at the outset of her studies it was said of her that she one day begged a coachman not to drive so fast, by the title of " spoon " : " Spoon, spoon, pray not so fast"; using the word *cucchiaio* instead of *cocchiere*.

The effect of this kind of mistake being in proportion to the gravity of the intention, I know of none better than that of an honest German (the late Mr. Stumpff, the harp-maker), who being disgusted at some trait of worldliness which he heard related, and wishing to say that rather than be guilty of such meanness he would quit society for a hermitage, and live upon acorns, exclaimed with great animation, "Oh, I shall go into de vilderness, and live upon *unicorns.*"

HUME AND THE THREE LITTLE KINGS.

When Hume was in Paris, receiving the homage of the philosophers for his skepticism, and of the courtiers for his advocacy of Charles I., three little boys were brought before him to make him speeches. They complimented him after the fashion of grown persons, said how impatiently they had expected his arrival, and expressed their admiration of his beautiful history. Alas! a history too much like that of the Stuarts was in preparation for them. These children were afterward the unfortunate Louis XVI., and his brothers Louis XVIII. and Charles X.

"Heaven from all creatures hides the Book of Fate,
All but the page prescribed—their present state."

If the poor little boys could have read in that tremendous volume, their compliments might have been turned in something of this fashion:

Little Charles X.—Accept the compliments,

Mr. Jacobite, of a prince whom you will help to send into exile.

Little Louis XVIII.—And of one whom you will help to bring from it, only to let him die of fat.

Little Louis XVI.—And of another, whose head your beautiful history will help to cut off.

A CHARMING CREATURE.

Shakespeare, in the compass of a line, has described a thoroughly charming girl :

" Pretty, and witty ; wild, and yet, too, gentle."

BACON.

If I were asked to describe Bacon as briefly as I could, I should say that he was the liberator of the hands of knowledge.

SUICIDES OF BUTLERS.

Tragedy will break in upon one's dinner-table in spite of us. Mr. Wakley tells us that suicide is rife among *butlers!* The news is startling to people at dinner. How many faces must have been turned on butlers, the day on which the coroner made the remark ; and how uncomfortable some of them must have felt ! The teetotalers will not overlook it ; for the cause appears obvious enough. The butler is always sipping.

He is also the most sedentary of domestics, the housekeeper excepted; and wine-merchants accuse him of having a bad conscience. So he grows burly and uneasy; thinks he shall never retire into an inn or a public office; loses bits of his property in speculation; and when the antibilious pill fails him, there is an inquest.

The poor butler should take to his legs instead of his arm-chair. He should make himself easier in his mind, considering his temptations; and cultivate an interest in everything out of doors, except shares in railroads.

DUELS.

The only conjecture to be made as to the possible utility of duels (on the assumption that the retention of any prevailing custom must have some foundation in reason) seems to be, that they serve to counteract the effeminate tendencies of sedentary states of society, and admonish us of the healthiness and necessity of courage.

For as to suffering insolence and outrage, the most polished nations of antiquity had no duels, and yet never appear to have felt the want of them.

But the Greeks and Romans, by their wrestling-grounds, and military training, and the very nakedness and beauty of their sculpture, maintained a sense of the desirableness of bodily vigor.

The diffusion of knowledge, however, seems

to be conspiring with the increased activity and practical good sense of the age to discountenance dueling, and render it ridiculous; and as the occasions of it are in general really so, while the consequences are tragical to the persons concerned, it is to be hoped that every brave and considerate man will do what he can to assist in proving it superfluous.

Did anybody ever write a serious panegyric on a duel? It has received hundreds of banters, and (consequences apart) has a natural tendency to the burlesque. Nay, even those have given rise to it in some pensive minds.

About thirty years ago, there was a famous duel about a couple of dogs between a Colonel Montgomery and a Captain Macnamara, in which the former was killed. The colonel or the captain would not "call his dog off," and the captain or colonel would not hinder his dog from going on; and so

"Straight they called for swords and pistols,"

and made a few women and children miserable.

This catastrophe occasioned a printed elegiac poem, the author of which, who was quite serious, concluded it with a burst of regret in the following extraordinary triplet:

"If two fine dogs had quarreled not!—Oh! *if*
Not fell Montgomery through false honor's *tiff*,
Nor Chalk-Farm witnessed of two heroes' *miff!*"

LISTON.

Talking of paralysis reminds one of the death of Liston. Poor fellow! he had long outlived the active portion of his faculties, and used to stand at his window by Hyde Park Corner, sadly gazing at the tide of human existence which was going by, and which he had once helped to enliven.

Liston's "face was his fortune." He was an actor, though truly comic and original, yet of no great variety; and often got credit given him for more humor than he intended, by reason of that irresistible compound of plainness and pretension, of chubbiness and challenge, of born, baggy, desponding heaviness, and the most ineffable airs and graces, which seemed at once to sport with and be superior to the permission which it gave itself to be laughed at. When Liston expressed a peremptory opinion, it was the most incredible thing in the world, it was so refuted by some accompanying glance, gesture, or posture of incompetency. When he smiled, his face simmered all over with a fondness of self-complacency amounting to dotage. Never had there been the owning of such a soft impeachment.

Liston was aware of his plainness, and allowed himself to turn it to account; but not, I suspect, without a supposed understanding between him and the audience as to the superiority of his intellectual pretensions; for, like many comedians, he was a grave man underneath his mirth, thought himself qualified to be a tragedian, and did, in fact, now and then act in tragedy for his benefit, with a lamentable sort of respectability that disappointed the laughers. I have seen him act in this way as Octavian in "The Mountaineers."

STEEPLE-CHASING.

Steeple-chasing is to proper bold riding what foolhardiness is to courage. It proves nothing except that the chaser is in want of a sensation, and that he has brains not so much worth taking care of as those of other men.

A. But is it not better than stag-hunting?

B. For the stag, certainly.

A. There can be no such piteous sight at a steeple-chase as may be seen at other kinds of hunting.

B. How can you be sure of that? I am afraid you are severer upon the chasers than I am.

A. Suppose, as the poet says,

"A stag comes weeping to a pool."

B. Good; but suppose

"A wife comes weeping to a fool."

Suppose Numskull brought home on a shutter. Danger for danger's sake is senseless. Besides, the horse is worth something. One has no right to crash and mash it in a pit on the other side of a wall, even with the chance of being retributively kicked to death in its company. Did you ever hear this patient and noble creature, the horse, scream for anguish? It is one of the ghastliest and most terrific of sounds ; one of the most tremendous even on a field of battle ; and depend upon it, you will catch no old soldier risking the chance of hearing it. If you do, he will be no Uncle Toby, nor Major Bath, nor the "Iron Duke" himself; but some brazen-faced simpleton, with no more brains in his head than his helmet.

TURKEYS.

It is amusing to see the turkey strutting and gobbling about the homestead. He looks like a burlesque on the peacock. Good old Admiral S. ! How sorry he was to hear the simile ; and what good things he had to say on the worth of turkeys in general, and of a foreign species of the race in particular. But is it not true ? Look at the animal's attempt to get up a sensation with his "tail," or what is called such. Look at the short-coming size of it, the uncouth heaviness of his body, the somber tawdriness of his colors, and, above all, that ineffable drawing back of the head and throat into an intensity of the arrogant and self-satisfied ! He looks like a corpulent fop in a paroxysm of conceit. John Reeve was not greater in the character of Marmaduke Magog the beadle, when he stamped the ground in a rapture of pomp and vanity. Bubb Dodington might have looked so when he first put on his peer's robes, and practiced dignity before a looking-glass. The name of Bubb is very turkey-like. The bird's familiar name in Scotland, admirably expressive of its appearance, is Bubbly Jock. Goethe says that Nature has a lurking sense of comedy in her, and sometimes intends to be jocose; and it is not difficult to imagine it when one considers that she includes art, and comedy itself, and is the inventress of turkeys.

The turkey is a native of America, and Franklin recommended it for the national symbol!

BAGPIPES.

An air played on the bagpipes, with that detestable, monotonous drone of theirs for the bass, is like a tune tied to a post.

CÆSAR AND BONAPARTE.

To-morrow (Sunday, the 15th of the month) is the famous Ides of March, the day of the death of Cæsar. During a conversation which Napoleon had with the German poet Wieland, he expressed his surprise at the "great blunder" of which Cæsar was guilty; and, on the poet intimating by his look a desire to know what the blunder was, his Majesty said, it was trusting people with his life whose designs against it he was aware of. Wieland thought within himself, as he contemplated the imperial countenance, " That is a mistake that will never be committed by you." But see how dangerous it is for a living man to pronounce judgment on a dead one. If Napoleon would never have committed the mistake of Cæsar, the accomplished Roman would not have fallen, as the other did, for want of knowing the character of the nations with whom he fought, and the chances of a climate. Now, it is better to perish in consequence of having a generous faith than a self-satisfied ignorance.

PSEUDO-CHRISTIANITY.

Some religious persons the other day, with a view to the promotion of " Christian union," had a meeting in Birmingham, at which they are said to have come to these two resolutions: First, that it is "*everybody's* right and duty to exercise private judgment in the interpretation of the Scriptures"; and second, that "*nobody* is to belong to their society who does not hold the doctrine of the divine institution of the Christian ministry, and the authority and perpetuity of Baptism and the Lord's Supper."

This is the way Christianity has been spoilt ever since dogma interfered with it—ever since something was put upon it that had nothing to do with it, in order that people might dictate to their neighbors instead of loving them, and indulge their pragmatical egotism at the very moment when they pretend to leave judgment free and to promote universal brotherhood. It is just as if some devil had said : " Christianity shall *not* succeed ; people shall not be of one accord, and find out what's best for 'em. I'll invent dogma ; I'll invent faith *versus* reason ; I'll invent the Emperor Constantine ; I'll invent councils, popes, polemics, Calvins and Bonners, inquisitions, auto-da-fés, massacres ; and should Christianity survive and outgrow these, I'll invent frights about them, and whispers in their favor, and little private popes of all sorts, all infallible, all fighting with one another, all armed with their *sine qua nons*, for the purpose of beating down the olive-branch, and preventing their pretended object from superseding my real one."

I do not believe, mind, that any such thing was said, or that this chaos of contradiction has been aught else but a fermentation of good and ill, out of which good is to come triumphant, perhaps the better for the trial ; for evil itself is but a form of the desire of good, sometimes a necessity for its attainment. But the seeming needlessness of so much evil, or for so long a period, is provoking to one's uncertainty ; and the sight of such a heap of folly is a trial of the patience.

Our patience we must not lose, for then we shall fall into the error we deprecate; but let us keep reason and honest ridicule for ever on the watch.

A. But they say that ridicule is unfair.

B. Yes; and make use of it whenever they can. In like manner they deprecate reason, and then reason in favor of the deprecation.

DYED HAIR.

There is a sly rogue of a fellow advertising in the Dublin papers, who is very eloquent and dehortatory on the subject of gray hair. He says that people, when they begin to have it, decline "in respect and esteem" as "companionable beings," particularly with the fair sex; nay, in their own eyes; and therefore he advises them to lose no time in availing themselves of an immense discovery which he has made, in the shape of a certain "coloring material," which turns the hair instantly to a "luxuriant dark." He tells them that it is as easy in the operation as combing, preserves and invigorates as well as beautifies, will not stain the most delicate linen, is useless for any other purpose, and in fine will not cost them a farthing. All they have to pay is "two pounds" for the secret. He does not quit his theme without repeating his caution as to the dreadful consequences that will ensue from neglecting his advice—that "decided change," as he calls it, "which a gray or a bald head is sure to produce in public, private, and self-esteem."

Every gentleman, not quite perfect in the color of his hair, must start at this advertisement, "like a guilty thing surprised." He must think of all the friends, particularly female ones, in whose eyes he might or ought to have noticed a manifest decrease of his acceptability; must begin to reflect how painful it is to lose caste as a "companionable being"; and what steps he ought to take, in order to recover his threatened standing in public and private estimation. "Good heaven!" he will exclaim, looking in the glass, "and is it come to this? I see it; I feel it. Yes, there is a 'decided change'; virtue is gone out of me. Miss Dickenson looks odd; Lady Charlotte is dignified; nobody will hold me in any further regard; perhaps I shall lose my office, my estate, my universe. I am a lost, middle-tinted man."

So saying, he disburses his two pounds in a frenzy, realizes the wonderful dark hair immediately; and, in the course of two days, what is the consequence? I remember an elderly gentleman whose sister persuaded him to adopt one of these two-pound secrets that cost nothing. All he had to do was to make use of a comb dipped in the preparation, and the fine dark color undoubtedly resulted. In the course of a few hours it changed to a beautiful *blue;* and he had the greatest difficulty, for days after, to get rid of it.

EATING.

Talk of indulgence in eating as you may, and avoid excess of it as we must, it is not a little wonderful to consider what respect Nature entertains for the process, and how doubly strange and monstrous the consideration renders the wants of the half-starved. It throws us back upon thoughts more amazing still. We observe that the vital principle in the universe, instead of, or perhaps in addition to, its embodying itself in the shapes of created multitudes throughout the apparently uninhabited portions of space, tends to concentrate its phenomena into distinct dwelling-places, or planets, in which they are so crowded together (though even then with large seeming intervals) that they are compelled to keep down the populations of one another by mutual devourment. Fortunately (so to speak, without meaning at all to assume that fortune settles the matter), this cruel-looking tendency is accompanied by Nature's usual beneficent tendency to produce a greater amount of pleasure than pain; for the duration of the act of dying, or of being killed, is in no instance comparable with that of the state of being alive; and life, upon the whole, is far more pleasurable than painful (otherwise we should not feel pain so impatiently when it comes). The swallow snaps up the fly; the fly has had its healthy pleasures; and one dish entertains at a time many human feasters. Now think of the enormous multitude of those dishes—of the endless varieties of *food* which Nature seems to have taken a delight in providing, and of the no less diversity of tastes and relishes with which she has recommended them to our palates. Take the list of eatables for mankind alone (if any cook could make one out), and think of its endless variety of fish, flesh, and fowl, of fruits, and vegetables, and minerals; how many domestic animals it includes; how many wild ones; how many creatures out of the sea; how many trees and shrubs; how many plants and herbs; how many lands, oceans, airs, climates, countries, besides the combinations producible out of all these results by the art of cookery (for art is also Nature's doing); modifications of roast and boiled and broiled, of pastries, jellies, creams, confections, essences, preserves. One would fancy that she intended us to do nothing but eat; and, indeed, a late philosopher said that her great law was, "Eat, or be eaten." The philosopher obeyed it pretty stoutly himself (it was Darwin), and he inculcated it (one would think with no great necessity) on his patients; some of whose biliary vessels must have contributed to pay him well for the advice.

For here is the puzzle. A man stands equally astonished at the multitude of his temptations to eat, at the penalty of the indulgence, and at the starvations of the poor. I am not going to enter into the question, or to endeavor to show how it may be reconciled with the beneficence of Nature in a large and final point of view, the only point in which her great operations can be regarded. What I meant to show was her respect for this eating law of hers, and the astonishing spirit of profusion in which she has poured forth materials for its exercise. Why we are not all individually rich or healthy enough to do it justice is another question, which can not, indeed, but suggest itself during the consideration. Mr. Malthus (as if that mended the matter) said there was not room enough to squeeze in at the table between himself and his bishop! Let us comfort ourselves (till the question is settled) by reflecting that the mortal portion of Mr. Malthus, and of the bishop too, have gone to nourish the earth which is to support the coming generations. "Fat be the gander" (as the poet says) "that feeds on their grave."

If you are ever at a loss to support a flagging conversation, introduce the subject of eating. Sir Robert Walpole's secret for unfailing and harmonious table-talk was gallantry; but this will not always do, especially as handled by the jovial minister. Even scandal will not be welcome to everybody. But who doesn't eat? And who can not speak of eating? The subject brightens the eyes and awakens the tender recollections of everybody at table—from the little boy with his beatific vision of dumpling, up to the most venerable person present, who mumbles his grouse. "He that will not mind his dinner," said Johnson, "will mind nothing" (he put it stronger; but honest words become vulgarized, and the respectable term "stomach" won't fit). Ask a lady if she is attached to the worthiest gentleman in the room, and she will reasonably think you insult her; but ask if she is "*fond of veal*," and she either enthusiastically assents, or expresses a sweet and timid doubt on the subject—an apologetical inability to accord with those who are. She "*can't say she is.*" "Love" was formerly the word; perhaps is still.

"Do you love pig?"

"No, I can't say I do; but I dote upon eels."

QUESTIONER (*looking enchanted*). "Really! Well, so do I."

Dishes are bonds, not only of present, but of absent unanimity. I remember an uxorious old gentleman, who had a pretty wife that he was recommending one day to the good graces of a lady at the head of a table. His wife was not present; but he had been expatiating on her merits, and saying how Mrs. Scrivelsby did this thing and did that, and what a charming, elegant woman she was, when the conversation became diverted to other topics, and the lady's accomplishments lost sight of. The gentleman's hostess happening to speak of some fish at table, he asked if she "loved the roe"; and upon her owning that "soft impeachment," and being helped to some, he exclaimed, in the fondest tones, with a face full of final bliss, and radiant with the thoughts of the two sympathizing women, the absent and the present—

"Do you, indeed? Well, now, Mrs. Scrivelsby loves the roe."

N. B.—If anybody sees "nothing" in this story, he is hereby informed that he has made a discovery unawares; for that is precisely the value of it.

POLAND AND KOSCIUSKO.

The claims of Poland may be imperfect. She was once badly governed; there is no doubt of that; but so are many nations who, nevertheless, very properly decline to be governed by others; and, besides, she has had bitter teaching, and professes to have learned by it. Her leaders are not so confined, as they are supposed to be, to the aristocracy. Kosciusko himself was no aristocrat—hardly, indeed, a Pole proper. He was a small gentleman of Lithuania; but he loved his half-countrymen, the Poles; and he thought, with Blake, that they ought not to be "fooled by foreigners."

One of the most affecting of national anecdotes is related of this great man during the first occupation of France by the Allies. He was then living there, but siding neither with the Allies nor with Bonaparte. He never did side with either. He knew both the parties too well. A Polish troop in the allied service came foraging in his neighborhood, and they took liberties with his humble garden. The owner came out of the house, and remonstrated with them in their own language.

"Who are you," said they, exasperated, "that are not on our side, and yet dare to speak to us in this manner?"

"My name is Kosciusko."

They fell at his feet, and worshiped him.

ENGLAND AND THE POPE (GREGORY).

The Pope, instead of attending to the welfare of the unfortunate people whom he governs, and saving his country from the reproach of being the worst governed state in Europe, is putting up prayers to Heaven for the conversion of England! He might as well come to London, and try to convert Mr. Cobden to the corn-laws, or the railway companies to the old roads.

About eighty years ago, a Scotsman went to Rome for the purpose of converting the Pope. The Scotsman was not content with praying. He boldly entered St. Peter's at high mass, and addressed his Holiness in a loud voice, by the title of a certain lady who lives not a hundred miles from Babylon. The Pope, who at that time, luckily for the Scotsman, happened to be a kind and sensible man (Ganganelli), was advised to send him to the galleys; but he answered that the galleys were but a sorry place to teach people "good breeding"; so he put the honest fanatic into a ship, and sent him home again to Scotland.

We, in England here, should be equally civil to the Pope, if he would do us the honor of a visit; and he might take Dr. Pusey away with him if he pleased, together with a score or two of ladies and gentlemen who constitute converted England.

It is a little too late in the day to expect Englishmen to pant after purgatory and confession; to rejoice in the damnation of their fathers and mothers and little children; or even to wish for the celibacy of their clergy. Their clergy are accused of being lively enough already toward the ladies. What would they be if they had no wives? "Gracious heavens!" Why, in the course of six months the bench of bishops would be as bad as cardinals.

THE DUKE OF WELLINGTON'S CONCERT.

The Duke of Wellington has been directing a concert of Ancient Music. It is curious to see the music he selected: what a mixture it is of devotion, fighting, and gallantry; how he abides by the favorites of his youth; and how pleasantly, and like a good son, he includes the compositions of his father, Lord Mornington. Conquerors deal in such tremendous (and disputable) wares, that it is not easy to determine the amount of their genius—to distinguish it from chances and consequences, or to say how much of it is not owing to negative as well as positive qualities. The world are hardly in a condition to judge a man who plays at chess with armies; who blows us up, takes us by storm and massacre, and alters the face of nations. He may or may not be as great as we suppose; though his want of civil talents is generally against him, and he often perishes out of imprudence. But there can be no doubt that a great soldier is a very striking and important person of some kind; and to catch him at these soft, harmonious, and filial amusements is interesting. The Duke's concert the other day was in good old taste, not omitting some of the later great masters. There was plenty of Handel in it; some Gluck and Paisiello, Beethoven and Mozart; Avison's "Sound the loud timbrel"; a glee of Webbe's; another by his Grace's father, aforesaid; and the fine old French air, "Charmante Gabrielle," which, an arch rogue of a critic says, was sung in a "chaste manner" by Madame Caradori. Not that the chastity is to be doubted, or that the air was not one of recognized propriety; but it is worth considering how

"Nice customs curtsey to great kings";

what storms of honor and glory, and royal and national trumpets, have been allowed to smuggle into good society the "Charming Gabrielle," mistress of Henri Quatre; and how the fair singer would have been scared at being requested to do as much for the charming Jane Shore, or giddy Mrs. Eleanor Gwynn.

The Duke of Wellington's was a right soldier's concert, a little overdone perhaps in the church-going quarter—a little too much on the oratorio side; but that might have been looked for at an "Ancient Concert," famous for memories of George the Third. The rest was given to love and fighting—to "Gird on thy Sword," and "Giving them Hailstones," and "Charming Gabrielle," and the ladies' duet in Mozart, "Prenderò quel brunettino" (I'll take that little brown fellow), which may have been connected with some pleasing reminiscences of country-quarters or the jungles of Hyderabad.

But the paternal glee, after all, was the thing; the filial reminiscence; the determination of the great "iron" Duke to stand by his little, gentle, accomplished father, the amateur composer—and a very pretty composer too. All soldiers can go to church and admire charming Gabrielles; but it is not for every great fame thus to stand by a minor one, and take a pride in showing off the father on whose knee it sat in its infancy.

The Duke is a good fellow, depend upon it,

me patre judice. He may give odd answers to deputations, and be "curst and brief" to autograph-seekers, and not know how to talk in their own language to his warm-hearted Irish countrymen. I wish he did. But he sticks to his father. He will have due honor paid to the paternal crotchets.

WAR, DINNER, AND THAKKSGIVING.

It is not creditable to a "thinking people" that the two things thay most thank God for should be eating and fighting. We say grace when we are going to cut up lamb and chicken, and when we have stuffed ourselves with both to an extent that an orang-outang would be ashamed of; and we offer up our best praises to the Creator for having blown and sabered his "images," our fellow creatures, to atoms, and drenched them in blood and dirt. This is odd. Strange that we should keep our most pious transports for the lowest of our appetites and the most melancholy of our necessities! that we should never be wrought up into paroxysms of holy gratitude but for bubble and squeak, or a good-sized massacre! that we should think it ridiculous to be asked to say grace for a concert or a flower-show, or the sight of a gallery of pictures, or any other of the divinest gifts of Heaven, yet hold it to be the most natural and exalted of impulses to fall on our knees for having kicked, beaten, torn, shattered, drowned, stifled, exenterated, mashed and abolished thousands of our "neighbors," whom we are directed to "love as ourselves"!

A correspondent of the "Times," who had of course been doing his duty in this respect, and thanking Heaven the first thing every morning for the carnage in the Punjaub, wished the other day to know "what amount of victory was considered, by the Church or State, to call forth a public expression of thankfulness to Almighty God." He was angry that the Bishops had not been up and stirring at the slaughter; that Sir Robert Peel was not as anxious to sing hymns for it as to feed the poor; that Lord John Russell, with all his piety, was slower to call for rejoicings over the Sikh widows than attention to hapless Ireland.

The pause did Government honor. The omission of the ceremony, if they had had courage enough to pass it by altogether, would have done them more. Not because God is not to be reverenced in storm as in sunshine, but because it does not become any section of his creatures to translate these puzzles of the mystery of evil in their own favor, and, with the presumptuous vanity called humility, thank Him, like the Pharisee, for not being conquered like "those" Indians. Our meddling with the Punjaub at all is connected with some awkward questions. So is our whole Indian history. I believe it to have been the inevitable, and therefore, in a large and final point of view, the justifiable and desirable consequence of that part of the "right of might" which constitutes the only final secret of the phrase, and which arises from superior knowledge and the healthy power of advancement. But in the humility becoming such doubtful things as human conclusions, it behooves us not to play the fop at every step; not to think it necessary to God's glory or satisfaction to give Him our "sweet voices," even though we do it in their most sneaking tones; nor to thank the good Father for having been chosen to be the scourgers of our weaker brethren.

"Go," we might imagine Him saying; "go, and hold your tongues, and be modest. Don't afflict me during the necessity with your stupid egotism. Perhaps I chose you for the task, only because you had the less sensibility."

FIRES AND MARTYRDOM.

Fires are still happening every day, notwithstanding the tremendous lessons which they give to the incautious. People are shocked at the moment, and say that something must be done; but in the course of four-and-twenty hours they forget the shrieking females at the windows, and the children reduced to ashes; and the calamities are risked as before. It is really a pity that Parliament does not interfere. Officious legislation is bad; but if the public are children in this respect, and don't know how to take care of themselves, grown understandings ought to help them. Parliament can ordain matters about lamps and pavements; why not about balconies for great houses, and corridors at the back of smaller ones? Are health and convenience of more importance than being saved from the cruelest of deaths?

Meantime, what an opportunity presents itself to Puseyites and others for a little indisputable Christianity—a good practical restitution of their favorite days of martyrdom and self-sacrifice. It is said that no calamitous chance of things is ever done away with in this country, unless some great man happens to be the victim. Now, the Puseyites are accused of being Christian only in disputation, with great dislikes of foregoing their comforts and snug corners. Here is an occasion for them to prove their brotherly love—to show how their gold can be tried in the fire. Why can not Dr. Pusey, or Mr. Newman, or Mr. Wells (who admires the tapers and other splendid shows of Popery) be a shining light himself, of the most unquestionable order? Why not take some house about to be pulled down in a great thoroughfare, assemble a crowd at night-time, set fire to the goods and chattels round about him like an Indian widow, step forth into the balcony to show us how easy it was for him to escape, and then, in spite of our cries, tears, agonies, and imploring remonstrances (the more, the memorabler), offer himself up, like a second Polycarp, on the altar of human good? Invidious people say, that it is no very difficult thing for a man to be a shining light in a good comfortable pulpit, between breakfast and dinner, with no greater heat on him than that of his self-complacency; but the Ridleys and Bradfords found a different business of it at the stake; and here is an opportunity for such as sneer at those Protestant martyrs, to show how they can be martyrs themselves of a nobler sort, and of the most undoubted utility. For who could forget the circumstance? what balconies and corridors would not start forth to their honor and glorification all over the metropolis?

But perhaps the Bishop of London, who is jealous of his prerogative, might choose to avail himself of the opportunity. Or suppose Bishop Philpotts requested it of him as a favor. What a truly reviving spectacle, in these days of Christian declension, to see the two bishops, at the last moment, affectionately contesting with one another the honor of the sacrifice, and trying to thrust his brother off the devoted premises!

RESPECTABILITY.

"When the question was put to one of the witnesses on the trial of Thurtell, 'What sort of a person was Mr. Weare?' the answer was, 'Mr. Weare was *respectable.*' On being pressed by the examining counsel as to what he meant by respectability, the definition of the witness was, that '*he kept a gig!*'"

"A person," says the "York Courant," on this incident, "was annoying a whole company in a public room, and one of them reproving him sharply for his indecorum, an apologist whispered: 'Pray, do not offend the gentleman; I assure you he is a *respectable* man. He is worth *two hundred a year independent property.*"

There is no getting at the root of these matters, unless we come to etymology. People mean something when they say a man is respectable; they mean something different from despicable or intolerable. What is it they do mean? Why, they mean that the gentleman is worth *twice looking at*—he is respectable, *re-spectabilis;* that is to say, literally, one who is to be looked at *again.* You must not pass him as though he were a common man; you must turn round and observe him well; a second look is necessary if you have the least respect for him; if you have more, you look at him again and again; and if he is very respectable indeed, and you have the soul of a footman, you look at him till he is out of sight, and turn away with an air as if you could black his shoes for him.

But what *is* "respectable"? What is the virtue that makes a man worth twice looking at? We have intimated it in what has been said. The "York Courant" has told us—*he keeps a gig.* Gig is virtue. A buggy announces moral worth. *Curriculus evehit ad deos.*

But you must be sure that he does keep it. He may come in a gig, and yet the gig not be his own; in which case it behooves you to be cautious. You must not be taken in by appearances. He may look like a gentleman; he may be decently dressed; you may have seen him perform a charitable action; he may be a soldier covered with scars, a patriot, a poet, a great philosopher; but for all this, beware how you are in too much haste to look twice at him—the gig may have been borrowed.

On the other hand, appearances must not condemn a man. A fellow (as you may feel inclined to call him) drives up to the door of an inn; his face (to your thinking) is equally destitute of sense and goodness; he is dressed in a slang manner, calls for his twentieth glass of gin, has flogged his horse till it is raw, and *condemns*, with energetic impartiality, the eyes of all present, his horse's, the bystanders', and his own. Now, before you pronounce this man a blackguard, or think him rather to be turned away from with loathing than looked at twice out of respect, behave you as impartially as he: take the ostler aside, or the red-faced fellow whom he has brought in the gig with him, and ask, "Is the gig his own?" The man, for aught you know, may reply: "His own? Lord love you, he has a mint of money. He could ride in his coach if he pleased. He has kept a gig and Moll Fist these two years." Thus you see, without knowing it, you might have loathed a respectable man. "He keeps his gig."

But this respectable gentleman not only keeps his gig—he might keep his coach. He is respectable *in esse;* *in posse* he is as respectable as a sheriff: you may look twice at him; nay, many times. Let us see. We have here a clew to the degrees of a man's respectability. To keep a gig is to be simply respectable: you may look twice at the gig-man. A curricle, having two horses, and costing more, is, of course, more respectable: you may look at the possessor of a curricle at least twice and a half. A chariot renders him fit to be regarded over and over again; a whole carriage demands that you should many times turn your neck to look at him; if you learn that he drives a coach and four, the neck may go backward and forward for three minutes; and if the gentleman abounds in coaches, his own carriage for himself, and another for his wife, together with gig, buggy, and dog-cart, you are bound to stand watching him all the way up Pall Mall, your head going like a fellow's jaws over a pan-pipe, and your neck becoming stiff with admiration.

The story of the "two hundred a year independent property" is a good appendage to that of the gig-keeping worthy. The possessor of this virtue was annoying a whole company in a public room, and one of them reproving him for his indecorum, somebody whispered: "Do not offend the gentleman; he is a respectable man, I assure you. He is worth two hundred a year independent property." The meaning of this is: "I am a slave, and believe you to be a slave: think what strutting fellows *we* should be if we possessed two hundred a year; and let us respect ourselves in the person of this bully."

If people of this description could translate the feelings they have toward the rich, such is the language their version would present to them, and it might teach them something which they are ignorant of at present. The pretense of some of them is, that money is a great means of good as well as evil, and that of course they should secure the good and avoid the evil. But this is not the real ground of their zeal; otherwise they would be zealous in behalf of health, temperance, and honesty, good-humor, fair dealing, generosity, sincerity, public virtue, and everything else that advances the good of mankind. No; it is the pure, blind love of power, and the craving of weakness to be filled with it. Allowance should be made for much of it, as it is the natural abuse in a country where the most obvious power is commercial; and the blindest love of power, after all (let them be told this secret for the comfort of human nature), is an instinct of sympathy—is founded on what others will think of us, and what means we shall find in our hands for adding to our importance. It is this value for one another's opinion which keeps abuses so long in existence; but it is in the same corner of the human heart, now that reform has begun, that the salvation of the world will be found.

USE OF THE WORD "ANGEL," ETC., IN LOVE-MAKING.

Lady Suffolk, when bantering Lord Peterborough on his fondness for the fine terms used in love-making, said that all she argued for was, that as these expressions had been in all ages the favorite words of fine gentlemen, who would persuade themselves and others that they are in love, those who really are in love should discard them, the better to distinguish themselves from impostors. But, with submission to her ladyship, a real lover may take them up again, as they were first taken up, because with him the language is still natural.

ELOQUENCE OF OMISSION.

A late gallant Irishman, who sometimes amused the House of Common and alarmed the Ministers with his *brusquerie* (Mr. Montague

Mathew, I believe), set an ingenious example to those who are at once forbidden to speak, and yet resolved to express their thoughts. There was a debate upon the treatment of Ireland, and the General, having been called to order for taking unseasonable notice of the enormities attributed to Government, spoke to the following effect: "Oh, very well; I shall say nothing then about the murders—(*Order, order!*)—I shall make no mention of the massacres—(*Hear, hear! Order!*)—Oh, well; I shall sink all allusion to the infamous half-hangings—(*Order, order! Chair!*)"

This Montague Mathew was the man who, being confounded on some occasion with Mr. Mathew Montague (a much softer-spoken gentleman), said, with great felicity, that people might as well confound "a chestnut horse with a horse-chestnut."

GODS OF HOMER AND LUCRETIUS.

Sir William Temple says that he "does not know why the account given by Lucretius of the gods should be thought more impious than that given by Homer, who makes them not only subject to all the weakest passions, but perpetually busy in all the worst or meanest actions of men." Perhaps the reason is, that in Homer they retain something of sympathy with others, however misdirected or perturbed; whereas the gods of Lucretius are a set of selfish *bons-vivants*, living by themselves and caring for nobody.

AN INVISIBLE RELIC.

Bruges is the place where the Catholics professed to have in their keeping the famous *hau de Saint Joseph;* that is to say, one of the *ho!'s* which St. Joseph used to utter when in the act of cleaving wood as a carpenter. The reader may think this a Protestant invention; but the story is true. Bayle mentions the *ho* in his Dictionary.

A NATURAL MISTAKE.

A little girl seeing it written over inn doors, "Good stabling and an ordinary on Sundays," thought that the stabling was good on week-days but only ordinary on the Sabbath.

MORTAL GOOD EFFECTS OF MATRIMONY.

A lady meeting a girl who had lately left her service, inquired, "Well, Mary! where do you live now?" "Please ma'am," answered the girl, "I don't live now—I'm married."

UMBRELLAS.

From passages in the celebrated verses of Swift on a "Shower," which appeared in 1770, and in Gay's poem of "Trivia, or the Art of walking the Streets," which was written a year or two afterward, it would seem that the use of umbrellas at that time was confined to females, and those too of the poorer classes. The ladies either rode in their carriages through the rain, or were obliged to fly from it into shops.

"Now in contiguous drops the flood comes down,
Threatening with deluge this devoted town.
To shops in crowds the draggled females fly,
Pretend to cheapen goods, but nothing buy.
The Templar spruce, while every spout's abroach,
Stays till 'tis fair, yet seems to call a coach.
The tucked-up seamstress walks with hasty strides,
While streams run down her oiled umbrella's sides."

There is no mention of an umbrella for men. The men got under a shed, like the Templar, into a coach, or into a sedan.

"Here various kinds, by various fortunes led,
Commence acquaintance underneath a shed;
Triumphant Tories and desponding Whigs
Forget their feuds, and join to save their wigs.
Boxed in a chair, the beau impatient sits,
While spouts run clattering o'er the roof by fits;
And ever and anon, with frightful din,
The leather sounds: he trembles from within.
So when Troy-chairmen bore the wooden steed,
Pregnant with Greeks, impatient to be freed
(Those bully Greeks, who, as the moderns do,
Instead of paying chairmen, run them through),
Laocoön struck the outside with his spear,
And each imprisoned hero quaked for fear."

In Gay's poem, the men are advised, in case the weather threatens rain, to put on their surtouts and *worst wigs*. The footman, he says, lets down the flat of his hat. Even among the females, the use of the umbrella appears to have been confined to winter-time :

> " Good housewives all the winter's rage despise,
> Defended by the riding-hood's disguise ;
> Or, underneath the umbrella's oily shed,
> Safe through the wet, on chinking pattens tread.
> Let Persian dames th' umbrella's ribs display,
> To guard their beauties from the sunny ray ;
> Or sweating slaves support the shady load,
> When Eastern monarchs show their state abroad :
> Britain *in winter only* knows its aid,
> To guard from chilly show'rs the walking maid."

When Jonas Hanway made his appearance with an umbrella, the vulgar hooted him for his effeminacy.

Umbrellas, it is observable, are always mentioned as being oiled. I think I remember the introduction of silken ones.

BOOKSELLERS' DEVICES.

Mr. Pickering, with no unpleasing pedantry, gives his edition of the Poets the epithet of " Aldine." Aldus was the great elegant publisher of his day, and Mr. Pickering is ambitious of being thought his follower. He adopts his device in the title-page, with a motto calculated to mystify the unlearned—"Aldi Discipulus Anglus"; to wit, Aldus's English Disciple. This is good, because anything is good that has faith in books or elegance of choice; but, inasmuch as originality is a good addition to it, a device of Mr. Pickering's own would have been better. Aldus's dolphin is very well done, but it is somewhat heavy.

Mr. Taylor, the printer, a man of liberal knowledge, has a device of his own—a hand pouring oil into the midnight lamp; and the late Mr. Valpy had another, not so good, a digamma (the Greek F), which looked like an improvement upon a gallows. It seemed as if it was intended to hang two commentators instead of one; or the parson, with his clerk underneath him.

WOMEN ON THE RIGHT SIDE.

Dr. A. Hunter said that women who love their husbands generally lie on their right side. What did he mean by "generally"? Women who love their husbands always lie on the right side, for an obvious reason—to wit, that they can not lie on the wrong one.

SHENSTONE MISTAKEN.

It is strange that Shenstone should have thought his name liable to no pun. A man might have convinced him to the contrary, after the fashion in which Johnson proposed to help forgetfulness. "Sir," said the Doctor to somebody who was complaining of short memory, "let me give you a kick on the shin, and I'll be bound you'll never forget it." So a man might have thrown a stone at Shenstone's leg and said, "There, Mr. Shinstone"; for, as to the *i* and the *e*, no punster stands upon ceremony with a vowel.

THE MARSEILLES HYMN.

The "Marseilles Hymn," though not in the very highest class of art, in which pure feeling supersedes the necessity of all literal expression, is nevertheless one of those genuine compositions, warm from the heart of a man of genius, which are qualified to please the highest of the scientific, and those who know nothing of music but by the effect it has upon them. The rise upon the word *Patrie* (or, as the English translator has very well made it fall, upon the word *Glory*) is a most elevating note of preparation; this no sooner rouses us to war, than we are reminded of the affecting necessity for it in the threats of the tyrants, followed by that touching passage respecting the tears and cries of our kindred; and then comes another exalting note—the call to arms. The beating of the drum succeeds. We fancy the hurried muster of the patriots; their arms are lifted, their swords unsheathed; and then comes the march—a truly grand movement—which even on the piano-forte suggests the fullness of a band. In the pathetic part, the E flat on the word *fils* and the whole strain on that passage are particularly affecting. The tears seem to come into the eyes of the heroes, as no doubt they have into thousands of them, and into thousands of those that have heard the song. But it must be played well, and not be judged of by the performance of a new or a feeble hand.

I know not who the author of the translation, or rather imitation, is, but he has done it very well.

NON-SEQUITUR.

There is a punning epigram by Dr. Donne which is false in its conclusion:

"'I am unable,' yonder beggar cries,
 'To stand or go.' If he says true, he lies."

No; because he may lean, or be held-up.

NON-RHYMES.

It is curious that in so correct a writer as Pope, and in so complete a poem as the "Rape of the Lock," there should be two instances of rhyme which are none at all:

"But this bold Lord, with manly strength en*dued*,
 She with one finger and a thumb sub*dued*."

> "The doubtful beam long nods from side to *side;*
> At length the wits mount up, the hairs sub*side.*"

They are both in the fifth canto. There is another in the "Essay on Criticism":

> "Unfinished things one knows not what to *call,*
> Their generation's so equivo*cal.*"

STOTHARD.

The death of Stothard grieved all the lovers of art, though it had been long expected. They regretted to think that they could have no "more last words" from his genius—no more of those sweet and graceful creations of youth, beauty, and womanhood, which never ceased to flow from his pencil, and which made his kindly nature the abode of a youthful spirit to the last. An angel dwelt in that tottering house, amid the wintery bowers of white locks, warming it to the last with summer fancies.

Stothard had the soul in him of a genuine painter. He was a designer, a colorist, a grouper; and, above all, he had expression. All that he wanted was a better education, for he was never quite sure of his drawing. The want was a great one; but, if those who most loudly objected to it had had a tenth part of his command over the human figure, or even of his knowledge of it (for the purposes of expression), they would have had ten times the right to venture upon criticising him; and, having that, they would have spoken of him with reverence. His class was not of the highest order, and yet it bordered upon the gentler portion of it, and partook of that portion; for, since the days of the great Italian painters, no man felt or expressed the graces of innocence and womanhood as he did. And his coloring (which was little known) had the true relish, such as it was. He loved it, and did not color for effect only. He had a bit of Rubens in him, and a bit of Raphael—and both of them genuine; not because he purposely imitated them, but because the seeds of gorgeousness and of grace were in his own mind. The glowing and sweet painter was made out of the loving and good-natured man. This is the only process. The artist, let him be of what sort he may, is only the man reflected on canvas. The good qualities and defects of his nature are there; and there they will be, let him deny or disguise them as he can. In youth, Stothard was probably too full of enjoyment, and had too little energy, to study properly. In the greater masters, enjoyment and energy, sensibility and strength of purpose, went together. Inferiority was the consequence; but inferiority only to *them*. The genius was indestructible.

Stothard, for many years, was lost sight of by the public, owing to the more conventional elegances of some clever but inferior men, and the dullness of public taste; but it was curious to see how he was welcomed back as the taste grew better, and people began to see with the eyes of his early patrons. The *variety*, as well as grace, of his productions soon put him at the head of designers for books, and there he remained. What he did for the poems of Mr. Rogers is well known, and his picture of the Canterbury Pilgrims still better, though it was not one of his best. Many of his early designs for "Robinson Crusoe" and other works, especially those in the old "Novelist's Magazine," far surpass it; and so do others in Bell's "British Poets." There is a female figure bending toward an angel, in one of the volumes of Chaucer in that edition, which Raphael himself might have put in his portfolio; and the same may be said of larger designs for editions of Milton and Shakespeare. See, in particular, those from "Comus," and for the "Two Gentlemen of Verona," where there is a girl in boy's clothes. Nothing can be more true or exquisite than the little doubtful gesture of fear and modesty in the latter figure, blushing at the chance of detection. Stothard excelled in catching these fugitive expressions of feeling—one of the rarest of all beauties. But he has left hundreds, perhaps thousands, of designs—rich treasures for the collector and the student. He is one of the few English artists esteemed on the Continent, where his productions are bought up like those of his friend Flaxman, who may be reckoned among his imitators; for Stothard's genius was richer than his, and included it.

THE COUNTENANCE AFTER DEATH.

A corpse seems as if it suddenly knew everything, and was profoundly at peace in consequence.

HUME.

Hume, the most unphilosophic (in some respects) of all philosophic historians, and a bigoted enemy of bigotry (that is to say, unable to give candid accounts of those whom he differed with on certain points), was a good-natured, easy man in personal intercourse, dispassionate, not ungenerous, and could do people kind and considerate services. Out of the pale of sentiment, and of what may be called the providential and possible, he was an unanswerable, or at least an unanswered dialectician; but there was a whole world in that region into which he had no insight; and for want of it he was not qualified to pronounce finally on matters of faith and religion.

GIBBON.

Gibbon was a skeptic, in some respects, of a similar kind with Hume, and more immersed in the senses. I say "more," because both these anti-spiritual philosophers were fat, double-chinned men. Perhaps Gibbon's life was altogether a little too selfish, and lapped up in cotton. He lumbered from his bed to his board, and back again, with his books in the intervals, or rather divided his time between the three, in a sort of swinishness of scholarship. Martyrdom and he were at a pretty distance! He was not a man to die of public spirit, or to comprehend very well those who did. But his skepticism tended to promote toleration. He was an admirable Latin scholar, a punctilious historian, an interesting writer in spite of a bad style; and his faults, of every kind, appear to have been owing to temperament and disease, and to his having been an indulged infant, and heir to an easy fortune. Let us be thankful we got so much out of him, and that so diseased a body got so much out of life. A writer's infirmities are sometimes a reader's gain. If Gibbon had not disliked so much to go out of doors, we might not have had the "Decline and Fall."

ANGELS AND FLOWERS.

It might be fancied that the younger portion of angels—the childhood of heaven—had had a part assigned them in the creation of the world, and that they made the flowers.

Linnæus, however, would have differed on this point.

AN ENVIABLE DISTRESS.

Mr. Rogers, according to the newspapers, has been robbed of plate by his footman to the amount of two thousand pounds. What a beautiful calamity for a poet! to be *able* to lose two thousand pounds!

SIR THOMAS DYOT.

The street lately called Dyot Street, in St. Giles's, is now christened (in defiance, we believe, of a legal proviso to the contrary) George Street. It is understood that Sir Thomas Dyot, an admirable good fellow in the reign of the Stuarts, left his property in this street for the use and resort of the houseless poor who "had not where to lay their heads," upon condition of its retaining his name; and how the parish authorities came to have a right to alter the name his admirers would *like to know.*

It is a singular instance of the effect of circumstances in human affairs, that a name so excellent, and worthy to be had in remembrance, should become infamous in connection with this very street; and perhaps the authorities might undertake to vindicate themselves on that score, and ask whether Sir Thomas could have calculated upon such a vicissitude? But I say he could, and very likely did; for he knew of what sort of people the houseless poor were likely to be composed; and he was prepared, like a thorough-going friend, to take all chances with them, and trust to more reflecting times to do justice both to him and to them.

Or, if he did not think of all this, his instinct did ; or, at all events, it did not care for anything but playing the kind and manly part, and letting a wise Providence do the rest. Sir Thomas was a right hearty good fellow, whoever he was ; for nothing else, I believe, is known of him ;—a little wild, perhaps, in his youth ; otherwise he might not have become acquainted with the wants of such people ; but ever, be sure, honest to the backbone, and a right gentleman—fit companion for the Dorsets and Cowleys in their old age, not for the Charles the Seconds. Here's a libation to him in this dip of ink—in default of a bumper of Burgundy.

ANCIENT AND MODERN EXAMPLE.

One has little sympathy after all with the virtues or failings of illustrious Greeks and Romans. One fancies that it was their business to be heroical, and to furnish examples for school-themes—owing, perhaps, to the formality and tiresomeness of those themes. We leave then the practice and glory of their virtues as things ancient and foreign to us, like their garments, or fit only to be immortalized in stone—petrifactions of ambitious ethics, not flesh and blood, or next-door neighbors; stars for the sky, not things of household warmth and comfort; not feasible virtues—or, if feasible, rendered alien somehow by distance and strangeness, and perhaps accompanied by vices which we are hardly sorry to meet with, and which our envy (and something better) converts into reconcilements of their virtue; as when we hear, for example, that old Cato drank, or that Phocion said an arrogant thing on "the hustings," or that Numa (as a Frenchman would say) visited a pretty girl "of afternoons" —Ma'amselle Egérie—who, he pretended, was a goddess and an oracle, and gave him thoughts on legislation. So, of the professed men of pleasure in the ancient world—or indeed of professed men of pleasure at any time (for their science makes them remote and peculiar, a sort of body apart, *excessively Free* Masons)—one doesn't think one's self bound to resemble them. Their example is not pernicious, much less of any use for the attainment of actual pleasure. Who thinks of imitating the vices of Cæsar or Alexander, out of an ambition of universality? (what a preposterous fop would he be!) or stopping to drink and carouse when he ought to be moving onward, because Hannibal did it? or of being a rake because Alcibiades had a reputation of that sort (unless, perhaps, it be some one of our lively ultra-classical neighbors, whose father has indiscreetly christened him *Alcibiade,* and who studies Greek beauty in a ballet)? We do not think of imitating men in Greek helmets or the Roman toga. Their example is only for school-exercises, or to be brought forward in the speech of some virgin orator. We must have heroism in a hat and boots, and good-fellowship at a modern table. It is our every-day names, Smith, Jones, and Robinson, that must be instanced for an example which we can thoroughly feel. Has Thompson done a handsome action? Everybody cries, "What a good fellow is Thompson!" Is a living man of wit effeminate and a luxurious liver? The example becomes perilous. It is no remote infection, no "Plague of Athens." The disease is next door—a pestilence that loungeth at noon—a dandy cholera.

Nobody cares much for Pætus and Arria, and the fine example they set. Those Romans seem bound to have set them, for the benefit of the "Selectæ é Profanis" and the publications of Mr. Valpy. Lucretia sits "alone in her glory," a kind of suicide-statue—too hard of example to be followed. We can not think, somehow, that she felt much, except as a personage who should one day be in the classical dictionaries. And Portia's appears an odd and unfeeling taste, who swallowed "burning coals," instead of having a proper womanly faint, and taking a glass of water.

But tell us of "Mrs. Corbet" (celebrated by Pope), who heroically endured the cancer that killed her, and we understand the thing. Recount us a common surgical case of a man who has his leg cut off without wincing; and, as we are no farther off than St. Bartholomew's Hospital, it comes home to us. Tell us what a good fellow Thomson the poet was, or how Quin took him out of a spunging-house with a hundred pounds, or how Johnson "loved to dine," or Cowper solaced his grief with flowers and verses, and we all comprehend the matter perfectly, and are incited to do likewise.

MILTON AND HIS PORTRAITS.

There can be little doubt that Milton, however estimable and noble at heart, was far from being perfect in his notions of household government. He exacted too much submission to be loved as he wished. His wife (which was a singular proceeding in the bride of a young poet) absented herself from him in less than a month after their marriage—that is to say, during the very honey-moon; and she staid away the whole summer with her relations. He made his daughters read to him in languages which they did not understand; and in one part of his works he piques himself, like Johnson, on being a good hater. Now, "good haters," as they call themselves, are sometimes very good men, and hate out of zeal for something they love; neither would we undervalue the services which *such* haters may have done mankind. They may have been necessary; though a true Christian philosophy proposes to supersede them, and certainly does not recommend them. But as all men have their faults, so these men are not apt to have the faults that are least disagreeable, even to one another; for it is observable that good haters are far from loving their brethren, the good haters on the other side; and their tempers are apt to be infirm and overbearing. In the most authentic portraits of Milton, venerate them as one must, it is difficult not to discern a certain uneasy austerity—a peevishness—a blight of something not sound in opinion and feeling.

WILLIAM HAY.

Hay, the author of an "Essay on Deformity," was a member of Parliament, and an adherent, but not a servile one, to the government of Sir Robert Walpole. He was author of several publications on moral and political subjects, interesting in their day, and not unworthy of being looked at by posterity. He was a very amiable and benevolent man, of which his essays afford abundant evidence; and his name is to be added to the list of those delightful individuals, not so rare as might be imagined, who surmount the disadvantages of personal exterior on the wings of beauty of spirit. It is observable, however, of these men, that they have generally fine eyes.

BISHOP CORBET.

It is related of this facetious prelate, who flourished in the time of Charles the First, and whose poems have survived in the collections, that, having been tumbled into the mud with a fat friend of his by the fall of a coach, he said that "Stubbins was up to the elbows in mud, and he was up to the elbows in Stubbins." During a confirmation, he said to the country people who were pressing too closely upon the ceremony, "Bear off there, or I'll confirm you with my staff." And another time, on a like occasion, having to lay his hand on the head of a very bald man, he turned to his chaplain and said, "Some dust, Lushington," to keep his hand from slipping.

Corbet's constitutional vivacity was so strong as hardly to have been compatible with episcopal decorum. But times and manners must be taken into consideration; and, though a bishop of this turn of mind would have been forced, had he lived now, to be more considerate in regard to times and places, there is no reason to doubt that he took himself for as good a churchman as he was an honest man. And liberties are sometimes taken by such men with serious objects of regard, not so much out of a light consideration, as from the confidence of love. Had Corbet lived in later times, he would, perhaps, have furnished as high an example of elegant episcopacy as any of the

Rundles or Shipleys. As it was, he was a sort of manly college-boy, who never grew old.

HOADLY.

Hoadly, the son of the Bishop, and author of "The Suspicious Husband," was a physician, and a good-natured, benevolent man. His play has been thought as profligate as those of Congreve; but there is an animal spirit in it, and a native undercurrent of good-feeling, very different from the sophistication of Congreve's fine ladies and gentlemen. Congreve writes like a rake upon system; Hoadly like a wild-hearted youth from school.

VOLTAIRE.

Perhaps Voltaire may be briefly, and not unjustly, characterized as the only man who ever obtained a place in the list of the greatest names of the earth by an aggregation of secondary abilities. He was the god of cleverness. To be sure, he was a very great wit.

HANDEL.

Handel was the Jupiter of music; nor is the title the less warranted from his including in his genius the most affecting tenderness as well as the most overpowering grandeur; for the father of gods and men was not only a thunderer, but a love-maker. Handel was the son of a physician, and, like Mozart, began composing for the public in his childhood. He was the grandest composer that is known to have existed, wielding, as it were, the choirs of heaven and earth together. Mozart said of him that he "struck you, whenever he pleased, with a thunderbolt." His hallelujahs open the heavens. He utters the word "Wonderful" as if all their trumpets spoke together. And then, when he comes to earth, to make love amid nymphs and shepherds (for the beauties of all religions found room in his breast), his strains drop milk and honey, and his love is the usefulness of the Golden Age. We see his Acis and Galatea, in their very songs, looking one another in the face in all the truth and mutual homage of the tenderest passion; and poor jealous Polyphemus stands in the background, blackening the scene with his gigantic despair. Christian meekness and suffering attain their last degree of pathos in "He shall feed his flock," and "He was despised and rejected." We see the blush on the smitten cheek, mingling with the hair.

Handel had a large, heavy person, and was occasionally vehement in his manners. He ate and drank too much (probably out of a false notion of supporting his excitement), and thus occasionally did harm to mind as well as body. But he was pious, generous, and independent, and, like all great geniuses, a most thorough lover of his art, making no compromises with its demands and its dignity for the sake of petty conveniences. There is often to be found a quaintness and stiffness in his style, owing to the fashion of the day; and he had not at his command the instrumentation of the present times, which no man would have turned to more overwhelming account: but what is sweet in his compositions is surpassed in sweetness by no other; and what is great, is greater than in any.

MONTAIGNE.

Montaigne's father, to create in him an equable turn of mind, used to have him waked during his infancy with a flute.

Montaigne was a philosopher of the material order, and as far-sighted perhaps that way as any man that ever lived, having that temperament, between jovial and melancholy, which is so favorable for seeing fair play to human nature; and his good-heartedness rendered him an enthusiastic friend, and a believer in the goodness of others, notwithstanding his insight into their follies and a good stock of his own; for he lived in a coarse and licentious age, of the freedoms of which he partook. But, for want of something more imaginative and spiritual in his genius, his perceptions stopped short of the very finest points, critical and philosophical. He knew little of the capabilities of the mind, out of the pale of its more manifest influences from the body; his taste in poetry was logical, not poetical; and he ventured upon openly despising romances ("Amadis de Gaul," etc.), which was hardly in keeping with the modesty of his motto, "*Que sçais-je?*" (What do I know?)

Montaigne, who loved his father's memory, rode out in a cloak which had belonged to him; and would say of it, that he seemed to feel *wrapped up* in his father ("*il me semble m'envelopper de lui*"). Some writers have sneered at this saying, and at the conclusions drawn from it respecting the amount of his filial affection; but it does him as much honor as anything he ever uttered. There is as much depth of feeling in it as vivacity of expression.

WALLER.

Pope said of Waller, that he would have been a better poet had he entertained less admiration of people in power. But surely it was the excess of that propensity which inspired him. He was naturally timid and servile; and poetry is the flower of a man's real nature, whatever it be, provided there be intellect and music enough to bring it to bear. Waller's very best pieces are those in praise of sovereign authority and of a disdainful mistress. He would not have sung Saccharissa so well had she favored him.

OTWAY.

Otway is the poet of sensual pathos; for, affecting as he sometimes is, he knows no way to the heart but through the senses. His very friendship, though enthusiastic, is violent, and has a smack of bullying. He was a man of generous temperament, spoilt by a profligate age. He seems to dress up a beauty in tears, only for the purpose of stimulating her wrongers.

RAPHAEL AND MICHAEL ANGELO.

The lovers of energy in its visible aspect think Michael Angelo the greatest artist that ever lived. Ariosto (in not one of his happiest compliments), punning upon his name, calls him

"Michel, più che mortal, Angiol divino."
(Michael, the more than man, Angel divine.)

Pursuing the allusion, it may be said that there is much of the same difference between him and Raphael as there is between their namesakes, the warlike archangel Michael, in "Paradise Lost," and Raphael, "the affable archangel." But surely Raphael, by a little exaggeration, could have done all that Michael Angelo did; whereas Michael Angelo could not have composed himself into the tranquil perfection of Raphael. Raphael's gods and sibyls are as truly grand as those of Buonarroti; while the latter, out of an instinct of inferiority in intellectual and moral grandeur, could not help eking out the power of *his* with something of a convulsive strength—an ostentation of muscle and attitude. His Jupiter was a Mars intellectualized. Raphael's was always Jupiter himself, needing nothing more, and including the strength of beauty in that of majesty, as true moral grandeur does in nature.*

WAX AND HONEY.

Wax-lights, though we are accustomed to overlook the fact, and rank them with ordinary commonplaces, are true fairy tapers—a white metamorphosis from the flowers, crowned with the most intangible of all visible mysteries, fire.

Then there is honey, which a Greek poet would have called the sister of wax—a thing as beautiful to eat as the other is to look upon; and beautiful to look upon, too. What two extraordinary substances to be made, by little winged creatures, out of roses and lilies! What a singular and lovely energy in nature to impel those little creatures thus to fetch out the sweet and elegant properties of the colored fragrances of

---

* Since making these remarks, I have seen the bust of a Susannah, which, if truly attributed to Michael Angelo, proves him to have been the master of a sweetness of expression inferior to no man. It is indeed the perfection of loveliness.

the gardens, and serve them up to us for food and light!—honey to eat, and waxen tapers to eat it by! What more graceful repast could be imagined on one of the fairy tables made by Vulcan, which moved of their own accord, and came gliding, when he wanted a luncheon, to the side of Apollo!—the honey golden as his lyre, and the wax fair as his shoulders. Depend upon it, he has eaten of it many a time, chatting with Hebe before some Olympian concert; and as he talked in an undertone, fervid as the bees, the bass-strings of his lyre murmured an accompaniment.

ASSOCIATIONS WITH SHAKESPEARE.

How naturally the idea of Shakespeare can be made to associate itself with anything which is worth mention! Take Christmas for instance: "Shakespeare and Christmas"; the two ideas fall as happily together as "wine and walnuts," or heart and soul. So you may put together "Shakespeare and May," or "Shakespeare and June," and twenty passages start into your memory about spring and violets. Or you may say, "Shakespeare and Love," and you are in the midst of a bevy of bright damsels, as sweet as rose-buds; or "Shakespeare and Death," and all graves, and thoughts of graves, are before you; or "Shakespeare and Life," and you have the whole world of youth, and spirit, and Hotspur, and life itself; or you may say even, "Shakespeare and Hate," and he will say all that can be said for hate, as well as against it, till you shall take Shylock himself into your Christian arms, and tears shall make you of one faith.

BAD GREAT MEN.

There have, undoubtedly, been bad great men; but, inasmuch as they were bad, they were not great. Their greatness was not entire. There was a great piece of it omitted. They had heads, legs, and arms; but they wanted hearts, and thus were not whole men.

CICERO.

This great Roman special pleader—the lawyer of antiquity, the child of the old age of Roman virtue, when words began to be taken for things—was the only man ever made great by vanity.

FLOWERS IN WINTER.

It is a charming sight to see China roses covering the front of a cottage in winter-time. It looks as if we need have no winter, as far as flowers are concerned; and, in fact, it is possible to have both a beautiful and a fragrant garden in January. There is a story in Boccaccio of a magician who conjured up a garden in winter-time. His magic consisted in his having a knowledge beyond his time; and magic pleasures, so to speak, await on all who choose to exercise knowledge after his fashion, and to realize what the progress of information and good taste may suggest.

Even a garden six feet wide is better than none. Now the possessor of such a garden might show his "magic" by making the most of it, and filling it with color.

CHARLES LAMB.

Lamb was a humanist, in the most universal sense of the term. His imagination was not great, and he also wanted sufficient heat and music to render his poetry as good as his prose; but as a prose writer, and within the wide circuit of humanity, no man ever took a more complete range than he. He had felt, thought, and suffered so much, that he literally had intolerance for nothing; and he never seemed to have it, but when he supposed the sympathies of men, who might have known better, to be imperfect. He was a wit and an observer of the first order, as far as the world around him was concerned, and society in its existing state; for, as to anything theoretical or transcendental, no man ever had less care for it, or less power. To take him out of habit and convention, however tolerant he was to those who could speculate beyond them, was to put him into an exhausted receiver, or to send him naked, shivering, and driven to shatters, through the regions of space and time. He was only at his ease in the old arms of humanity; and humanity loved and comforted him like one of its wisest though weakest children. His life had experienced great and peculiar sorrows; but he kept up a balance between those and his consolations, by the goodness of his heart, and the ever-willing sociality of his humor; though, now and then, as if he would cram into one moment the spleen of years, he would throw out a startling and morbid subject for reflection, perhaps in no better shape than a pun, for he was a great punster. It was a levity that relieved the gravity of his thoughts and kept them from falling too heavily earthward.

Lamb was under the middle size, and of fragile make, but with a head as fine as if it had been carved on purpose. He had a very weak stomach. Three glasses of wine would put him in as lively a condition as can only be wrought in some men by as many bottles; which subjected him to mistakes on the part of the inconsiderate.

Lamb's essays, especially those collected under the signature of ELIA, will take their place among the daintiest productions of English *wit-melancholy*—an amiable melancholy being the groundwork of them, and serving to throw out their delicate flowers of wit and character with the greater nicety. Nor will they be liked the less for a sprinkle of old language, which was natural in him by reason of his great love of the old English writers. Shakespeare himself might have read them, and Hamlet have quoted them.

SPORTING.

The second of September is terrible in the annals of the French Revolution, for a massacre, the perpetrators of which were called *Septembrizers*. If the birds had the settlement of almanacs, new and startling would be the list of *Septembrizers* and their fusillades; amazing the multitude of good-humored and respectable faces that would have to look in the glass of a compulsory self-knowledge, and recognize themselves for slaughterers by wholesale—or worse, distributors of broken bones and festering dislocations.

"And what" (a reader may ask) "would be the good of that, if these gentlemen are not aware of their enormities? Would it be doing anything but substituting one pain for another, and setting men's minds upon needless considerations of the pain which exists in the universe?"

Yes; for these gentlemen are perhaps not quite so innocent and unconscious as, in the gratuitousness of your philosophy, you are willing to suppose them. Besides, should they cease to give pain, they would cease to feel it in its relation to themselves; and as to the pain existing in the universe, people in general are not likely to feel it too much, especially the healthy; nor ought anybody to do so, in a feeble sense, as long as he does what he can to diminish it, and trusts the rest to Providence and futurity. What we are incited by our own thoughts, or those of others, to amend, it becomes us to consider to that end: what we can not contribute any amendment to, we must think as well of as we can contrive. Sportsmen, for the most part, are not a very thoughtful generation. No harm would be done them by putting a little more consideration into their heads. On the other hand, all sportsmen are not so comfortable in their reflections as their gayety gives out; and the moment a man finds a contradiction in himself between his amusements and his humanity, it is a signal that he should give them up. He will otherwise be hurting his nature in other respects, as well as in this; he will be exasperating his ideas of his fellow creatures, of the world, of God himself; and thus he will be inflicting pain on all sides, for the sake of tearing out of it a doubtful pleasure.

"But it is effeminate to think too much of pain, under any circumstances."

Yes—*including* that of leaving off a favorite pastime.

Oh, we need not want noble pains, if we are desirous of them—pains of honorable endeavors, pains of generous sympathy, pains, most masculine pains, of self-denial. Are not these more manly, more anti-effeminate, than playing with life and suffering, like spoilt children, and cracking the legs of partridges?

Most excellent men have there been, and doubtless are, among sportsmen; truly gallant natures, reflecting ones too; men of fine wit and genius, and kind as mother's milk in all things but this—in all things but killing mothers because they are no better than birds, and leaving the young to starve in the nest, and strewing the brakes with agonies of feathered wounds. If I sometimes presume to think myself capable of teaching them better, it is only upon points of this nature, and because, for want of early habit and example, my prejudices have not been enlisted against my reflection. Most thankfully would I receive the wisdom they might be able to give me on all other points. But see what habit can do with the best natures, and how inferior ones may sometimes be put upon a superior ground of knowledge from the absence of it. Gilbert Wakefield I take to have been a man of crabbed nature, as well as confined understanding, compared with Fox; yet in the public argument which he had with the statesman on the subject, Wakefield had the best of it, poorly as it was managed by him. The good-natured legislator could only retreat into vague generalities and smiling admissions, and hope that his correspondent would not think ill of him. And who does? We love Fox always, almost when he is on the instant of pointing his gun; and we are equally inclined to quarrel with the tone and manner of his disputant, even when in the act of abasing it. But what does this prove, except the danger of a bad habit to the self-reconciling instincts of a fine enjoying nature, and to the example which flows from it into so much reconcilement to others? When a common hard-minded sportsman takes up his fowling-piece he is to be regarded only as a kind of wild beast on two legs, pursuing innocently his natural propensities, and about to seek his prey, as a ferret does, or a wild cat; but the more of a man he is, the more bewildered and dangerous become one's thoughts respecting the meeting of extremes. When Fox takes up the death-tube, we sophisticate for his sake, and are in hazard of becoming *effeminate* on the subject, purely to shut our eyes to the cruelty in it, and to let the pleasant gentleman have his way.

As to the counter-arguments about Providence and permission of evil, they are edge tools which it is nothing but presumption to play with. What the mind may discover in those quarters of speculation, it is impossible to assert; but, as far as it has looked yet, nothing is ascertained, except that the circle of God's privileges is one thing, and that of man's another. If we knew all about pain and evil, and their necessities, and their consequences, we might have a right to inflict them, or to leave them untouched; but not being possessed of this knowledge, and on the other hand being gifted with doubts, and sympathies, and consciences, after our human fashion, we must give our fellow creatures the benefit of those doubts and consciences, and cease to assume the rights of gods, upon pain of becoming less than men.

WISDOM OF THE HEAD AND OF THE HEART.

The greatest intellects ought not to rank at the top of their species, any more than the means rank above the end. The instinctive wisdom of the heart can *realize*, while the all-mooting subtlety of the head is only doubting. It is a beautiful feature in the angelical hierarchy of the Jews that the Seraphs rank first and the Cherubs after; that is to say, Love before Knowledge.

MÆCENAS

Wielded the Roman Empire with rings on his little finger. He deserves his immortality as a patron of genius; and yet he was a dandy of the most luxurious description amid the iron and marble of old Rome—the most effeminate of the effeminate, as Ney was "bravest of the brave." The secret of this weakness in a great man (for great he was, both as a statesman and a discerner of greatness in others) was to be found in excessive weakness of constitution.

LORD SHAFTESBURY'S EXPERIENCE OF MATRIMONY.

Shaftesbury was an honest man and politician, an elegant but fastidious writer, and, though a poor critic in poetry, could discern and forcibly expose the errors of superstition. In one of his letters is an extraordinary passage, not much calculated to delight the lady whom he married. He said he found marriage "*not so much worse*" than celibacy as he had expected! He appears to have had but a sorry *physique*.

A PHILOSOPHER THROWN FROM HIS HORSE.

Montaigne was one day thrown with great violence from his horse. He was horribly knocked and bruised within an inch of his life; was cast into a swoon; underwent agonies in recovering from it; and all this he noted down, as it were, in the faint light, the torn and battered tablets of his memory, during the affliction; drawing them forth afterward for the benefit of the reflecting. If you had met such a man in the streets, being carried along on a shutter, he would have been providing, as well as he was able, for your instruction and entertainment! This is philosophy, surely.

WORLDS OF DIFFERENT PEOPLE.

"The world!" The man of fashion means St. James's by it; the mere man of trade means the Exchange and a good prudent mistrust. But men of sense and imagination, whether in the world of fashion or trade, who use the eyes and faculties which God has given them, mean His beautiful planet, gorgeous with sunset, lovely with green fields, magnificent with mountains— a great rolling energy, full of health, love, and hope, and fortitude, and endeavor. Compare this world with the others. The men of fashion's is no better than a billiard-ball; the money-getter's than a musty *plum*.

MRS. SIDDONS

Was a person more admirable than charming, and not even so *perfectly* admirable on the stage as the prevalence of an artificial style of acting in her time induced her worshipers to suppose. She was a grand and effective actress, never at a loss, and equal to any demands of the loftier parts of passion; but her grandeur was rather of the queenlike and conventional order, than of the truly heroical. There was a lofty spirit in it, but a spirit not too lofty to take stage-dignity for the top of its mark. Mrs. Siddons was born and bred up in the profession, one of a family of actors, and the daughter of a mother of austere manners.

Mr. Campbell, in his Life of her, somewhat quaintly called her "the Great Woman"; but I know not in what respect she was particularly great as to womanhood. It was *queenhood*, not womanhood, that was her forte—professional greatness; not that aggregation of gentle and generous qualities, that union of the sexually charming and the dutifully noble, which makes up the idea of perfection in the woman.

Great women belong to history and to self-sacrifice, not to the annals of a stage, however dignified. Godiva gives us the idea of a great woman. So does Edward I.'s queen, who sucked the poison out of his arm. So does Abelard's Héloïse, loving with all her sex's fondness as long as she could, and able, for another's sake, to renounce the pleasures of love for the worship of the sentiment. Pasta, with her fine, simple manner and genial person, may be supposed the representative of a great woman. The greatness is relative to the womanhood. It only partakes that of the man, inasmuch as it carries to its height what is gentle and enduring in both sexes. The moment we recognize anything of what is understood by the word *masculine* in a woman (not in the circumstances into which she is thrown, but in herself or aspect), her greatness, in point of womanhood, is impaired. She should hereafter, as Macbeth says, "bring forth men-children only." Mrs. Siddons's extraordinary theory about Lady

Macbeth (that she was a fragile little being, very feminine to look at) was an instinct to this effect, repellent of the association of ideas which people would form betwixt her and her personation of the character.

Mrs. Siddons's refinement was not on a par with her loftiness. I remember in the famous sleeping-scene in "Macbeth," when she washed her hands and could not get the blood off, she made "a face" in passing them under her nose, as if she perceived *a foul scent.* Now, she ought to have shuddered and looked in despair, as recognizing *the stain on her soul.**

NON-NECESSITY OF GOOD WORDS TO MUSIC.

Music is an art that in its union with words in general may reasonably take, I think, the higher place, inferior as it is to poetry in the abstract. For when music is singing, the finest part of our senses takes the place of the more definite intellect, and nothing surely can surpass the power of an affecting and enchanting air in awakening the very flower of emotion. On this account, I can well understand a startling saying attributed to the great Mozart, that he did not care for having good words to his music. He wanted only the

---

* This trait of character has been mentioned in my "Autobiography"; but I leave it standing, partly for the sake of completing a sketch.

names (as it were) of the passions. His own poetry supplied the rest.

GOETHE.

If I may judge of Goethe from the beautiful translations of him by Shelley, Carlyle, Anster, and others, he had a subtile and sovereign imagination, was a master in criticism, was humane, universal, reconciling, a noble casuist, a genuine asserter of first principles, wise in his generation, and yet possessing the wisdom of the children of light. Nevertheless, it is a question whether any man daring to think and speculate as he has done, would have been treated with so much indulgence, if worldly power had not taken him under its wing, and had he not shown too conventional a taste for remaining there, and falling in with one of its most favored opinions. Goethe maintained that the great point for society to strain at was not to advance (in the popular sense of that word), but to be content with their existing condition, and to labor contentedly every man in his vocation. His conclusion, I think, is refuted by the simple fact of the existence of hope and endeavor in the nature of men. If society is determined never to be satisfied, still it will hope to be so; the hope itself may, for aught that can be affirmed to the contrary, be a mere part of the work—of the necessary impulse to activity; but there it is—now working harder than ever—and a thousand Goethes can not destroy, though they may daunt it. They must destroy hope itself first, and life, and death too (which is continually renewing the ranks of the hopeful and the young), and above all the press, which will never stop till it has shaken the world more even.

It was easy for a man in Goethe's position to recommend people to be content with their own. But to be content with some positions is to be superior to them; and yet Goethe after all, in his own person, was neither superior to, nor content with, the conventionalities which he found made for him. He did not marry the woman he lived with till circumstances, as he thought, compelled him; and this was late in life. And instead of being superior to his condition, as he recommended the poor and struggling to be, his very acquiescence in other conventionalities showed how little he was so. If this great universalist proved his superiority by condescension, it was at any rate by contracting his wings and his views into the court circle, and feathering an agreeable nest which he never gave up. Unluckily for the reputation of his impartiality, all his worldly advantages were on the side of his theory. It is, therefore, impossible to show that it was anything else but a convenient acquiescence. He hazarded nothing to prove it otherwise; though, in the instance of his non-marriage, he showed how willing he was to depart from it where the hazard was not too great. In England, he would have married sooner, or departed from his acquiescences more.

Goethe, on account of this opinion of his, and the position which he occupied, is not popular at present in Germany. The partisans of advance there do not like him, perhaps from a secret feeling that they are more theoretical than practical themselves, and that in this respect he has represented his native country too well. For honest Germany, perhaps because she is more material than she supposes, and has unwittingly acquired a number of charities and domesticities from a certain sensual *bonhomie*, which has given her more to say for herself in that matter than she or her transcendentalists would like to own, is far more contemplative than active in her politics, and willing enough to let other nations play the game of advancement, as long as she can eat, drink, and dream, without any very violent interruption to her self-complacency. Pleasant and harmless may she live, with beau ideals (and very respectable ones they are) in the novels of Augustus La Fontaine; and may no worse fate befall the rest of the world, if it is to get no farther. Much of it has not got half so far. Her great poet, who partook of the same *bonhomie* to an extent which he would have thought unbecoming his dignity to confess, even as a partaker of good things, "let the cat out of the bag" in this matter a little too ingenuously; and for this, and the court airs they thought he gave himself, his countrymen will not forgive him. It is easy for his wholesale admirers, especially for the great understandings among them (Mr. Carlyle, for instance), to draw upon all the possibilities of an abstract philosophy, and give a superfine unworldly reason for whatever he did; but we must take even great poets as we find them. Shakespeare himself did not escape the infection of a sort of livery servitude among the great (for actors were but a little above that condition in his time). With all his humanity, he finds it difficult to repress a certain tendency to browbeat the people from behind the chairs of his patrons; and though Goethe, living in a freer age, seldom indulges in this scornful mood (for it seems he is not free from it), yet it is impossible to help giving a little scorn for scorn, or at least smile for smile, when we see the poetical minister of state, with his inexperience of half the ills of life, his birth, his money, his strength, beauty, prosperity, and a star on each breast of his coat, informing us with a sort of patriarchal dandyism, or as Bonaparte used to harangue from his throne, that he is contented with the condition of his subjects and his own—"*France et moi*"—and that we have nothing to do but to be good people and cobblers, and content ourselves with a thousandth part of what it would distress him to miss.

BACON AND JAMES THE FIRST.

Bacon, in the exordium of his "Advancement of Learning," has expressed so much astonishment at the talents of King James the First (considering that he was "not only a king, but a king born"), that the panegyric has been suspected to be a "bold irony." I am inclined to think otherwise. Bacon was a born courtier, as well as philosopher; and even his philosophy, especially in a man of his turn of mind, might have found subtle reasons for venerating a being who was in possession of a good portion of the power of this earth.

GOLDSMITH'S LIFE OF BEAU NASH.

Nash is to be added to the list of long livers; and it is worthy of notice that what has been invariably observed of long livers, and appears (with temperance or great exercise) to be the only invariable condition of their longevity, has not failed in his instance: he was an early riser.

It has been doubted whether Goldsmith was the author of the "Life" attributed to him. I think, however, it bears strong internal marks of his hand, though not in its happiest or most confident moments. Its pleasantry is uneasy and overdone, as if conscious of having got into company unfit for it; and something of the tawdriness of the subject sticks to him—perhaps from a secret tendency of his own to mix up the external character of the fine gentleman, "in a blossom-colored coat," with his natural character as a writer. Chalmers, the compiler of the "Biographical Dictionary," who was much in the secrets of book-making, appears to have had no doubt on the subject. It is not improbable that Goldsmith had materials for the "Life," by some other person, put into his hands, and so made it up by touches of his own, and by altering the composition.

JULIUS CÆSAR.

Cæsar was one of the greatest men that ever lived, as far as a man's greatness can be estimated from his soldiership, and general talents, and personal aggrandizement. He had the height of genius in the active sense, and was not without it in the contemplative. He was a captain, a writer, a pleader, a man of the world, all in the largest as well as most trivial points of view, and superior to all scruples, except those which tended to the enlargement of his fame, such as clemency in conquest. Whether he was a very great man in the prospective, universal, and most enduring sense, as a man of his species, instead of a man of his time, is another question, which must be settled by the growing lights of the world and by future ages. He put an end to his country's freedom, and did no good, that I am aware of, to any one but himself, unless by the production or prevention of results known only to Providence.

FÉNELON.

Fénelon was a marvel of a man—a courtier yet independent, a teacher of royalty who really did teach, a liberal devotee, a saint in polite life. His "Telemachus" is not a fine poem, as some call it, but it is a beautiful moral novel. He had the courage to advise Louis XIV. not to marry the bigot Maintenon; and such was the respect borne to his character by the Duke of Marlborough and the other allied generals, that they exempted his lands at Cambray from pillage, when in possession of that part of Flanders. The utmost fault that could be found with him was, that perhaps the vanity attributed to Frenchmen found some last means of getting into a corner of his nature, in the shape of an over-studiousness of the feelings of others, and an apostolical humility of submission to the religious censures of the Pope. Charming blights, to be sure, in the character of a Catholic priest. The famous Lord Peterborough said of him, in his lively manner: "He was a delicious creature. I was obliged to get away from him, or he would have made me pious."

SPENSER AND THE MONTH OF AUGUST.

The word August deserves to have the accent taken off the first syllable and thrown upon the second (Augùst), not because the month was named after Augùstus (and yet he had a good deal of poetry in him too, considering he was a man of the world; his friend Virgil gives him even a redeeming link with the seasons), but because the month is truly an augùst month; that is to say, *increasing in splendor* till it fills its orb—majestic, ample, of princely beneficence—clothed with harvest as with a garment, full-faced in heaven with its moon.

Spenser, in his procession of the months, has painted him from a thick and lustrous palette:

> " The sixt was August, being rich arrayed
> *In garment all of gold, downe to the ground.*"

How true the garment is made by the familiar words " all of gold " ! and with what a masterly feeling of power, luxuriance, and music, the accent is thrown on the word " down " ! Let nobody read a great poet's verses either in a trivial or affected manner, but with earnest yet deliberate love, dwelling on every beauty as he goes. And pray let him very much respect his stops:

> " In garment all of gold;—*downe* to the ground.

> " Yet rode he not, but led a lovely maid
> *Forth* by the *lily hand*, the which was crowned
> With ears of corn;—and full her hand was found."

Here is a presentation for you, beyond all the presentations at court—August, in his magnificent drapery of cloth of gold, issuing forth, and presenting to earth and skies his Maiden with the lily hand, the highest bred of all the daughters of Heaven—Justice. For so the poet continues:

"That was the righteous Virgin, which of old
 Liv'd here on earth, and plenty made abound;
 But after Wrong was lov'd, and Justice sold,
 She left th' unrighteous earth, and was to heav'n extoll'd."

Extolled; that is, in the learned, literal sense, *raised out of*—taken away out of a sphere unworthy of her. *Ex*, out of; and *tollo*, to lift.

Many of Spenser's quaintest words are full of this learned beauty, triumphing over the difficulty of rhyme; nay, forcing the obstacle to yield it a double measure of significance, as we see in the instance before us; for the praise given to Justice is here implied, as well as the fact of her apotheosis. She is, by means of one word, *extolled* in the literal sense, that is to say, *raised up;* and she is *extolled* in the metaphorical sense, which means, praised and hymned.

ADVICE.

The great secret of giving advice successfully is to mix up with it something that implies a real consciousness of the adviser's own defects, and as much as possible of an acknowledgment of the other party's merits. Most advisers sink both the one and the other; and hence the failure which they meet with, and deserve.

ECLIPSES, HUMAN BEINGS, AND THE LOWER CREATION.

I once noticed a circumstance during an eclipse of the sun, which afforded a striking instance of the difference between humankind and the lower animal creation. The eclipse was so great (it was in the year 1820) that night-time seemed coming on; birds went to roost; and, on its clearing away, the cocks crew as if it was morning. At the height of the darkness, while all the people in the neighborhood were looking at the sun, I cast my eyes on some cattle in a meadow, and they were all as intently bent with their faces to the earth, feeding. They knew no more of the sun than if there had been no such thing in existence.

Two reflections struck me on this occasion: First, what a comment it was on the remarks of Sallust and Ovid, as to the prone appetites of brutes (*obedientia ventri*) and the heavenward privilege of the eyes of man (*cœlum tueri*); and, second (as a corrective to the pride of that reflection), how probable it was that there were things within the sphere of our own world of which humankind were as unaware as the cattle, for want of still finer perceptions; things, too, that might settle worlds of mistake at a glance, and undo some of our gravest, perhaps absurdest, conclusions.

This second reflection comes to nothing, except as a lesson of modesty. Not so the fine lines of the poet, which are an endless pleasure. How grand they are!—

>"Pronaque cum spectent animalia cætera terram,
>   Os homini sublime dedit, cœlumque tueri
>   Jussit, et erectos ad sidera tollere vultus."

Even Dryden's translation falls short, except in one epithet suggested by his creed:

>"Thus, while the mute creation downward bend
>   Their sight, and to their earthly mother tend,
>   Man looks aloft, and with erected eyes
>   Beholds his own hereditary skies."

This is good; and the last line is noble, both in structure and idea; but the phrase "man looks aloft," simple and strong as it is, is not so fine as man gifted with the "sublime countenance"; and "hereditary skies" conveys a modern belief not true to the meaning. The Pagans, you know, believed that men went into their heaven downward—into Elysium. "The Maker," says Ovid, "gave man *a sublime countenance*" (that is to say, in both senses of the word, "elevated"; for we must here take the literal and metaphorical meaning together), "and *bade* him contemplate the sky, and *lift his erected visage toward the stars.*"

Do not read, with some editions, "cœlumque

videre," which means to "see," and nothing more; but "cœlumque *tueri*," which means to see with "intuition"—with the mind.

EASTER-DAY AND THE SUN, AND ENGLISH POETRY.

It was once a popular belief, and a very pretty one, that the sun danced on Easter-day. Suckling alludes to it in his famous ballad:

> "Her feet beneath her petticoat,
> Like little mice stole in and out,
>   As if they feared the light;
> But, oh! she dances such a way,
> No sun upon an Easter-day
>   Is half so fine a sight."

It is a pity that we have, if not more such beliefs, yet not more such poetry to stand us instead of them. Our poetry, like ourselves, has too little animal spirits. It has plenty of thought and imagination; plenty of night-thoughts, and day-thoughts too; and in its dramatic circle a world of action and character. It is a poetry of the highest order and the greatest abundance. But, though not somber—though manly, hearty, and even luxuriant—it is certainly not a very joyous poetry. And the same may be said of our literature in general. You do not conceive the writers to have been cheerful men. They often recommend cheerfulness, but rather as a good and sensible practice than as something which they feel themselves. They have plenty of wit and humor, but more as satirists and observers than merry fellows. Addison was stiff, Swift unhappy, Chaucer always looking on the ground.

The fault is national, and therefore it may be supposed that we have no great desire to mend it. Such pleasure as may be wanting we take out in sulks.

But at more reasonable moments, or over our wine, when the blood moves with a vivacity more southern, we would fain see the want supplied—fain have a little more Farquhar, and Steele, and Tristram Shandy.

Cast your eyes down any list of English writers, such, for instance, as that at the end of Mr. Craik's "History of our Literature," and almost the only names that strike you as belonging to personally cheerful men are Beaumont and Fletcher, Suckling, Fielding, Farquhar, Steele, O'Keefe, Andrew Marvell, and Sterne. That Shakespeare was cheerful I have no doubt, for he was almost everything; but still it is not his predominant characteristic, which is thought. Sheridan could "set the table in a roar," but it was a flustered one at somebody's expense. His wit wanted good nature. Prior has a smart air, like his cap; but he was a rake who became cynical. He wrote a poem in the character of Solomon, on the vanity of all things. Few writers make you laugh more than Peter Pindar, but there was a spice of the blackguard in him. You could not be sure of his truth or his good will.

After all, it is not necessary to be cheerful in order to give a great deal of delight ; nor would the cheerfulest men interest us as they do if they were incapable of sympathizing with melancholy. I am only speaking of the rarity of a certain kind of sunshine in our literature, and expressing a natural, rainy-day wish that we had a little more of it. It ought to be collected. There should be a joyous set of elegant extracts—a " Literatura Hilaris " or " Gaudens "—in a score of volumes, that we could have at hand, like a cellaret of good wine, against April or November weather. Fielding should be the port, and Farquhar the champagne, and Sterne the malmsey ; and whenever the possessor cast an eye on his stock he should know that he had a choice draught for himself after a disappointment, or for a friend after dinner— some cordial extract of Parson Adams, or Plume, or Uncle Toby, generous as heart could desire, and as wholesome for it as laughter for the lungs.

THE FIVE-POUND NOTE AND THE GENTLEMAN.

It is a curious evidence of the meeting of extremes, and of the all-searching eyes of those tremendous luminaries the daily papers, that a man nowadays can not commit the shabbiest action in a corner, or hug himself never so much upon his cunning and privacy, but the next morning he shall stand a good chance of having it blazoned to the world. An instance occurred the other day. The porter of a house in Conduit Street picked up a five-pound note. A gentleman met him, who asked if he had seen such a thing. He said he had, gave it up, and was thanked with "a nod." The gentleman, retracing his steps, was accompanied awhile by the porter; and the latter, mustering up his courage, inquired if he did not think the circumstance worth a pot of beer. The gentleman (for this, his title, is judiciously repeated by the newspaper) made no other reply than by walking off to the other side of the street, "evidently satisfied," says the account, "that he was nothing out of pocket by losing his five-pound note."

If this man did not see the porter pick up the note, he is one of the shabbiest fellows on record; and if he did, he might as well have given him something in the gayety of his heart, if only by way of showing that all was right on both sides.

But was he able to give anything? Could he find it in his heart to disburse the fourpence? Was it within the compass of his volition? For, after one's first feeling of disgust, a poor devil like this, who can not say his groat's his own, has a right to a humane consideration. People are apt to imagine that anybody who has fourpence to spare, has nothing to do but to put his hand in his pocket and give it.

B. So he can, if he chooses.

A. Ay; but he can't choose.

B. Can't choose; oh, that is a phrase. You don't mean to say it literally?

A. Yes, I do. He is literally unable to choose. He can not choose if he would. The assertion is odd, and seems not very provable; but it may be illustrated, and proved too, I think, in a manner easy enough. Suppose a man has a paralysis of the arm, and can not lift it? You request him to lift it; but he can not do so. He is physically unable. Morally, he wishes to do it; he would choose it; he thinks himself a poor creature for the inability; but the act is out of his power. Now, there are cases in which the moral power is in a like miserable condition. Victims of opium have been known to be unable to will themselves out of the chair in which they were sitting; and victims of miserliness, in like manner, may be unable to will a penny out of their pockets. Their volition has a paralysis; and they can no more stir a finger of it than your man with the paralytic arm.

PAISIELLO.

Paisiello was one of the most beautiful melodists in the world, as the airs of "La Rachelina" and "Io sono Lindoro" would be sufficient to testify, if he had left us none of all his others. Those two are well known to the English public under the titles of "Whither, my Love," and "For Tenderness formed." But they who wish to know how far a few single notes can go in reaching the depths of the heart, should hear the song of poor Nina, "Il mio ben," in the opera of "Nina pazza per Amore." The truth and beauty of passion can not go further.

I admire the rich accompaniments of the Germans; but more accompaniment than the author has given to that song would be like hanging an embroidered robe on the shoulders of Ophelia.

CARDINAL ALBERONI.

Alberoni was the son of a gardener, and lived to the age of eighty-seven, sound in his faculties to the last. He said a thing remarkable for its address and fine taste; nobler, indeed, than he was probably aware of; and a lesson of the very highest theosophy. He was a man of vehement temper, as well as open discourse, and told a boy one day, who said he feared something, that he should "fear nothing, not even God himself."

The company looking shocked and astonished to hear such words from the mouth of a cardinal, Alberoni added, with a meek air and a softened voice, "For we are to feel nothing toward the good God but *love*."

SIR WILLIAM PETTY THE STATIST AND MECHANICAL PHILOSOPHER.

Sir William Petty was the son of a clothier, and was founder of the wealth, perhaps of the talent, of the Lansdowne family, who bear his name—their ancestor, the Earl of Kerry, having married his daughter. Sir William was a sort of Admirable Crichton in money-making; and he left a curious account of his accomplishments that way. Aubrey, who knew him, says that he had at one time been a shop-boy; and that while he was studying physic at Paris he was driven to such straits for a subsistence that "he lived a week or two on threepennyworth of walnuts."

Sir William was a physician, a surveyor, a member of Parliament, a timber-merchant, a political writer, a speculator in iron-works, fisheries, and lead-mines; and he wrote Latin verses, and was an active Fellow of the Royal Society. But for the particulars of his money-getting see his will, which is a curious specimen of a man of his sort, not always such a perfection of human wisdom as he seems to have supposed, but admirable for ingenuity and perseverance. He also appears to have been a wag and a buffoon! He "will preach extempore incomparably," says Aubrey, "either in the Presbyterian way, Independent, Capuchin friar, or Jesuit."

The same writer tells a pleasant story of him:

"Sir Thirom Sankey, one of Oliver Cromwell's knights, challenged Petty to fight with him. Petty was extremely short-sighted, and, being the challengee, it belonged to him to nominate place and weapon. He nominated, accordingly, a dark cellar and a carpenter's axe. This," says Aubrey, "turned the knight's challenge into ridicule, and it came to naught."

NAME OF LINNÆUS.

Linnæus's father was a clergyman, of a family of peasants. The customs of Sweden were so primitive at that time that people under the rank of nobility had no surnames; and, by a sort of prophetic inclination, the family of *Linnæus* had designated themselves from a favorite *linden* or *lime-tree*, which grew near their abode; so that Carl von Linné meant *Charles of the Lime-tree*. The lime was not unworthy of being his godfather.

JOHN BUNCLE (THE HERO OF THE BOOK SO CALLED).

Buncle is a most strange mixture of vehement Unitarianism in faith, liberality in ordinary judgment, and jovial selfishness in practice. He is a liberal, bigoted, whimsical, lawful sensualist. A series of *good fortunes* of a very peculiar description (that is to say, the loss of seven wives in succession!) enables him to be a kind of innocent Henry VIII. He argues a lady into the sacred condition of marriage, spends a delightful season with her, she dies in the very nick of time, and he tries as hard as he can to grieve for a while, in order that he may justify himself all the sooner in taking another. This is the regular process for the whole seven! With amazing animal spirits, iron strength, little imagination, and a relishing gusto, he is an amusing and lively narrator, without interesting our sympathy in the least, except in the relish with which he eats, drinks, and makes matrimony.*

POUSSIN.

Poussin, like Corneille, was a Norman. The addition of the earnest and grave character of the Normans to the general French vivacity rendered him one of the great names in art, fit to be mentioned with those of Italy. He had learning, luxuriousness, and sentiment, and gave himself up to each, as his subject inclined him, though never perhaps without a strong consciousness of the art as well as the nature of what he had to do. His historical performances are his driest; his poetical subjects full of gusto; his landscapes remote, meditative, and often with a fine darkness in them, as if his trees were older than any other

* The reader can see, if he pleases, more about this extraordinary person in the "Book for a Corner."

painter's. Shade is upon them, as light is upon Claude's.

Poussin was a genuine enthusiast, to whom his art was his wealth, whether it made him rich or not. He got as much money as he wanted, and would not hurry and degrade his genius to get more.

A pleasant anecdote is related of him, at a time when he must have been in very moderate circumstances. He spent the greatest part of his life at Rome, and Bishop (afterward Cardinal) Mancini being attended by him one evening to the door, for want of a servant, the Bishop said, "I pity you, Monsieur Poussin, for having no servant." "And I pity your lordship," said the painter, "for having so many."

The Bishop, by the way, must have been a very ill-bred or stupid man, to make such a remark. Fancy how beautifully Bishop Rundle, or Berkeley, or Thirlwall would *not* have said it! What respect they would have contrived to show to the non-possessor of the servant, without in the smallest degree alluding to the non-possession!

Was there no Roman Duke of Devonshire in those days, to teach men of quality how to behave?

PRIOR.

Prior wrote one truly loving verse, if no other. It is in his "Solomon." The monarch is speaking of a female slave, who had a real affection for him :

"*And when I called another, Abra came.*"

BURKE AND PAINE.

Paine had not the refinements which a nice education and a lively fancy had given to Burke. He could not discern, as his celebrated antagonist did, "the soul of goodness in things evil"—a noble faculty, when evil is to be made the best of. But the other's refinements, actuated by his vanity, led him to uphold the evil itself, because he could talk finely about it, and because others had undertaken to put it down without his leave. Self-reference and personal importance are at the bottom of everything that men do, when they do not show themselves ready to make sacrifices to the public good. If the vanity still remains the same in many, even when they do, it may be pardoned them as an infirmity which does not interfere with their usefulness. Burke began with being a reformer, and remained one as long as he drew attention to himself by it, and could command the respect of the "gentilities" among which he moved. When he saw, in contradiction to his prophecies, that the reform was to move in a wider sphere, and that he and his gentilities were not necessary to it, he was offended; turned right round to the opposite side; and wrote a book which George III. said every *gentleman* ought to read. " There was a time," says Paine, " when it was impossible to make Mr. Burke believe there would be any revolution in France. His opinion then was, that the French had neither spirit to undertake nor fortitude to support it ; and now that there is one, he seeks an escape by condemning it."

The first French Revolution was defaced by those actions of popular violence which were the result of a madness caused by the madness of the aristocracy. The foolish system of hostility to France in which Englishmen suffered themselves to be brought up by those who thought themselves interested in preserving it, easily allowed them to confound the evil with the good, and consequently to think ill of its advocates. Paine, therefore, was thought to write on a vulgar and pernicious side, while Burke had all the éclat of the gentilities.

The most vulgar thing which Paine did was to deny the utility of a knowledge of the dead languages. He had none himself ; and he saw the knowledge often vaunted by men who, having nothing else to boast of, possessed of course (in the proper sense of the word) not even that. He paid these men the involuntary compliment of showing them that *his* ignorance of the matter and *theirs* were pretty much on a par ; and as they exalted what they did not understand, he decried what he was ignorant of. It was a piece of inverted aristocracy in him—a privilege of non-possession.

THE DUTCH AT THE CAPE.

It is amusing to read of the ponderous indolence and cow-like ruminations of the Dutch settlers at the Cape of Good Hope. What an admirable word for them is "Settlers"!

Madame de Staël has given a ludicrous picture of the stiffness and formality of an English tea-table. Now, a Dutch tea-table is an English one cast in lead.

RUSSIAN-HORN BAND.

This, to be sure, is sounding the very "bass-note of humility." A man converted into a crotchet! An A flat in the sixtieth year of his age! A fellow creature of Alfred and Epaminondas, who has passed his life in acting a semitone! in waiting for his turn to exist, and then seizing the desperate instant, and being a puff!*

* "The Russian-horn music" (says an authority whose name I have forgotten) "was invented by Prince Gallitzin, in 1762. This instrument consists of forty persons, *whose life is spent in blowing one note.* The sounds produced are precisely similar to those of an immense organ, with this difference, that each note seems to blend with its preceding and following one —a circumstance which causes a blunt sensation to the ear, and gives a monotony to the whole. However, the effect pos-

DOGS AND THEIR MASTERS.

Mr. Jesse, in his "Anecdotes of Dogs," takes pains to prove that the dog is a better man than himself; but, love dogs as we may, we must not blaspheme their master. Dogs have admirable qualities. They possess, in particular, a most affecting and superabundant measure of attachment, of lovingness for their human companion—singular as regards the differences of the two beings, and wonderfully and beautifully superior to

sesses much sublimity when the performers are *unseen;* but, when they are visible, it is impossible to silence reflections which jar with their harmony. To see human nature *reduced to such a use* calls up thoughts very inimical to our admiration of strains so awakened. I inquired who the *instrument* belonged to (by that word both pipes and men are included), and was told it had just been purchased by a nobleman, on the recent death of its possessor. [They were serfs.]

"The band consists of twenty-five individuals, who play upon about fifty-five horns, all formed of brass of a conical shape, with the mouthpiece bent: the lowest of these horns is eight feet long and nine inches in diameter at the larger end, and sounds double A; the highest, which sounds E, is two inches and a half long, by one in diameter. Some of the horns, but not all, have keys for one or two semitones. When playing, the band is drawn up four deep, the trebles in front, and the very low horns laid on trestles at the back, so that the performer can raise the mouthpiece with ease, while the other end rests on the frame; one man plays the three lowest horns, blowing them in turns as they are wanted. Not the least curious portion of this machinery is the conductor, who, with the score before him on a desk, stands fronting his troops at what the common notions of self-interest : for, as Mr. Jesse's book shows, they are capable of quite as much attachment to the poorest as to the richest man, and, in the midst of the most hard-working and painful existence, will think themselves amply repaid by a crust and a caress. Delightful, admirable, noble, is the loving, hard-working, unbribable, martyr-spirited creature called the dog; who will die rather than desert his master under the most trying circumstances; who often does die, and (so to speak) breaks his heart for him, refusing to forsake his dead body or his grave.

soldiers would call the left flank of the company, and continues during the whole performance to beat the time *audibly* by tapping a little stick or cane on his desk. And this time he beats, not according to the equal divisions in a bar, but the number and quality of notes therein: thus, for a bar of three-fourth time, containing one crotchet followed by four quavers, he makes five taps, the last four twice as rapid as the first.

"When the performance began, notwithstanding all we had read, although we knew that each demisemiquaver of a rapid octave must be breathed by a separate individual, we were astonished at the unity of effect and correctness of time; and this feeling continued undiminished to the end. But to this our pleasure and approbation were confined, and, all moral considerations apart, we soon began to feel regret and pain that so much labor should have been bestowed on forming what may, probably, be a very first-rate band of Russian horns, but what is certainly a very second-rate band of wind-instruments. There is no expression, no coloring in the performance; and, though the tone produced by the bass horns is extremely fine and powerful, and the tenors are soft and mellow, the trebles are shrill, and very frequently sadly out of tune."

But still he must not be compared with the equally loving, more tried, and more awful creature called Man, with his conflicting thoughts, his greater temptations, his "looking before and after"; his subjection, by reason of his very superiority, to the most distressing doubts, fears, distracting interests, manifold ties, impressions of this world and the next—imaginations, consciences, responsibilities, tears. Between the noblest and most affectionate dog that dies out of a habit of love for his master, and the many-thoughted, many-hearted human being, who, loving existence and his family, can yet voluntarily face the gulf of futurity for some noble purpose, there is as much difference as between a thoughtless impulse and a motive burdened with the greatest drawbacks.

Thus much for the idle sentences quoted from Monsieur Blaze, Lord Byron, and others, about the superiority of dogs to men; things written in moments of spleen or ill will, contradicted by the writers in other passages, and thoughtlessly echoed, out of partiality to his subject, by kindly Mr. Jesse.

BODY AND MIND.

Pascal, in spite of his wisdom, was a victim to hypochondria and superstition. He was an admirable mathematician, reasoner, wit, and a most excellent man; and yet, notwithstanding this union of the most solid and brilliant qualities, a wretched constitution sometimes reduced him to a state which idiots might have pitied. As if his body had not been ill-treated enough, he wore an iron girdle with points on it next his skin; and he was in the habit of striking this girdle with his elbow, when a thought which he regarded as sinful or vain came across him. During his latter days he imagined that he saw an abyss by the side of his chair, and that he was in danger of falling into it. How modest it becomes the ablest men to be, and thankful for a healthier state of blood, when they see one of the greatest of minds thus miserably treated by the case it lived in!

WANT OF IMAGINATION IN THE COMFORTABLE.

People in general have too little imagination, and habit does not tend to improve it. The comfort, therefore, which they have derived ever since they were born from sustenance and warmth, they come to identify with the habitual feelings of everybody; and, though they read in the newspapers of the want of bread and fuel among the poor, it is with the utmost difficulty, and by a violent forcing of the reflection, that they can draw a distinction between the sensations of the poor man's flesh and stomach and those of their own. Hunger with themselves is brief; they can soon satisfy it. Cold is brief : they can go to the fire. They become unable to sympathize with the continuous operation of want. They think the poor man talks about cold and hunger, and that there must be some reason in it, inasmuch as he looks ill ; for they can picture to their imaginations a care-worn face, since they see so many about them, where the hands are warm and the stomachs well fed. But still, as their own hands are in the habit of being warm, and their stomachs comfortable, or at any rate uncomfortable with fullness, they have no abiding conception of hands cold for a whole day, or of an habitual craving for food.

I do not mean to say that it would be desirable for people to be over-sensitive on these points; otherwise the distress of half a dozen of human beings would be sufficient to discomfort the whole globe. It is to be hoped that the martyrs to reform and imagination will have suffered enough eventually to secure the infinite preponderance of good in this world. But, meanwhile, its advance is the slower for it, and the apathy of the excessively comfortable sometimes not a little provoking.

Take one of the clergymen, for instance, who have been writing addresses of late to the poor, to advise them to bear hunger and cold with patience. One of these gentlemen sits down to his writing-table, with his feet on a rug, before a good fire, after an excellent breakfast, to recommend to others the endurance of evils, the least part of which would rouse him into a remonstrance with his cook or his coal-merchant, perhaps destroy his temper, and put him in a state of un-Christian folly. "Bless me!" cries he, looking about him, if there is the least bit of a "crick" in the window, "how intolerably cold it is this morning!" and he rises from his chair, and not without indignation closes the intolerable window which the servant had so "shamefully neglected." His dinner is not ready when he returns from his ride. "'Tis very shameful of the cook," quoth he; "I have eaten nothing to signify since breakfast, and am ready to sink." The dinner is brought in with all trepidation, and he does sink—that is to say, into an easy-chair; and fish, flesh, and fowl sink into *him*. Little does he think, and less does he endeavor to think (for the thought is not a comfortable one), that the men to whom he wrote his address in the morning are in the habit of feeling this sinking sensation from morning till night, and of seeing their little crying children suffering from a distress which they *know to be so wretched.* Many of these poor people sink into the grave; and the comfortable clergyman thinks it much if he gets into his carriage, or puts his warm greatcoat and handkerchief round his portly neck, and goes to smooth the poor man's passage to that better world which he himself will keep aloof from as long as port and pheasant can help him.

"What riches give us, let us then inquire:
  Meat, fire, and clothes. What more? Meat, clothes, and fire."

These are the three great necessaries of life, meaning by meat, food. After a few lines to show the insufficiency of superfluities for rendering bad men happy, the poet says of these superfluities:

" Perhaps you think the poor might have their part.
  Bond damns the poor, and hates them from his heart.
  The grave Sir Gilbert holds it for a rule,
  That every man in want is knave or fool.

"' God can not love (says Blunt, with tearless eyes)
  The wretch he starves '—and piously denies.
  But the good Bishop, with a meeker air,
  Admits, *and leaves them,* Providence's care."

THE SINGING MAN KEPT BY THE BIRDS.

Want of Imagination plays strange tricks with most people. I will tell you a fable.

A traveler came into an unknown country where the people were more like birds than men, and twice as tall as the largest ostriches. They had beaks and wings, and lived in gigantic nests, upon trees of a proportionate size. The traveler, who was unfortunately a capital singer, happened to be indulging in one of his favorite songs, when he was overheard by a party of this monstrous people, who caught him and carried him home. Here he led such a life as made him a thousand times wish for death. The bird family did not seem to be cruel to one another, or even intentionally so to him; for they soon found out what he liked to eat, and gave him plenty of it. They also flattened him a corner of the nest for a bed; and were very particular in keeping out of his way a pet tiger which threw him into the most dreadful agitations. But in all other respects, whether out of cruelty or fondness, or want of thought, they teased him to death. His habitation, at best, was totally unfit for him. His health depended upon exercise, particularly as he was a traveler; but he could not take any in the nest, because it was hollow like a basin; and, had he attempted to step out of it, he would have broken his neck. Sometimes they would handle him in their great claws, till his heart beat as if it would come through his ribs. Sometimes they kissed and fondled him with their horrid beaks. Sometimes they pulled his nose this way and that, till he gaped and cried out for anguish; upon which they would grin from ear to ear, and stroke back his head, till the hairs came out by the roots. If he did not sing, they would pull his arms about, and cruelly spread out his fingers, as if to discover what was the matter with him; and, when he did sing to beguile his sorrows, he had the mortification of finding that they looked upon it as a mark of his contentment and happiness. They would sing themselves (for some of them were pretty good singing-birds for so coarse a species), to challenge him, as it were, to new efforts. At length our poor traveler fell sick of a mortal distemper, the termination of which was luckily hastened by the modes they took to cure it. "Wretch that I am!" cried he, in his last moments, "I used to think it unmanly to care about keeping a goldfinch, or even a lark; but all my manliness, in a like situation, can not prevent me from dying of torture."

A STRANGE HEAVEN.

I have often thought (don't be frightened) that if any one set of men ought to go to heaven more than another, it's rascals! Consider what fools the poor fellows are; what frights they undergo; what infamy they get; what ends they often come to; and, in most cases, what "births, parentages, and educations" they must have had. Or, if their anxieties have not been in proportion to their rascality, then consider what it is to want the feelings of other men; what bad pleasures it betrays them into, and of what good ones it deprives them. Think of those miserable dogs among them who have never even succeeded *as* rascals. Fancy Dick Dreary in his old age, toothless, despised, diseased, dejected, conscious that he has been all in the wrong, and unable to pay for a bit of fire in the winter to comfort his petty-larceny fingers. Is he to have nothing for all this? Oh, depend upon it that, if he has not had it already, of some unaccountable sort or other (which brings matters round), your rascal must come right somehow.

B. Theologians have various ways for settling that.

A. Yes, but not for all; and positively one single poor devil must not be omitted—no, not even though he be a Calvinist or an Inquisitor. Heavenly notions of justice are not to be at the mercy of the most infernal stupidity of mind. If I were a preacher, my doctrines would not go to flatter the poor dogs into crime with notions of certain kinds of absolution, which in that case it would be doubly infernal to refuse them. I should treat them as the fools which no men like to be called ; but, at the same time, as the pitiable fools which such men undoubtedly are.

Grave Gentleman. But a positive heaven for rascals!

A. (laughing). Oh, oh, *verbum sat.* Dante has heavens for his rascals—heavens even for the Emperor Constantine and the slayer of the Albigenses. Why mayn't we find a little blushing corner or so for Muggins, and Father Rack, and poor Dick Dreary?

STANDING GODFATHER.

To stand godfather is, I know, reckoned a very trifling ceremony: people ask it of others, either to gratify their own vanity or that of the person asked; they think nothing of the Heaven they are about to invoke. It is looked upon as a mere gossiping entertainment: a few child's squalls, a few mumbled amens, and a few mumbled cakes, and a few smirks accompanied by a few fees, and it is all over. The character and the peculiar faith of the promisers have nothing to do with it; the child's interest has nothing to do with it: the person most benefited is the parson, who is thinking all the time what sort of a present he shall get. Now, observe what I must do, should I undertake to be a godfather. I must come into the presence of God—a presence not to be slighted though in a private room—to worship Him with a falsehood in my mouth: that is, to make Him a profession of faith which I do not understand. I must then promise Him to teach the child this very faith which I do not understand, and to guard her youth from evil ways; when it is very probable I shall never be with her or see her, and most likely, if I did see her, I should get my head broken by her relations for giving impertinent advice. Considered in itself, I think the idea of christening a child, and answering for what one can not possibly foresee, a very ridiculous one;

but, when Heaven is called upon and the presence of the Deity invoked to witness it, it becomes a serious ceremony, though it may be an erroneous one; and the invocation of the Deity is not to be sported with even on an erroneous occasion.

MAGNIFYING TRIFLES.

Affection, like melancholy, magnifies trifles: but the magnifying of the one is like looking through a telescope at heavenly objects; that of the other, like enlarging monsters with a microscope.

RELICS.

It is amusing to think how the world neglect great men, and how they value their most paltry memorials; and yet it shows the happy tendency of every trifle to keep up their reputation. Thus the warrior who is ungratefully used by his country may obtain his reward after death by his cap or his sword; a poet may be immortalized among the vulgar by the chair in which he used to write; and the beautiful Mary Stuart triumph over her rival Elizabeth by the mere force of a miniature. Sometimes, indeed, this deification of kickshaws may be abused. The Roman Catholics have five or six legs, *original* legs, of the same saint, in five or six different places, so that either five of the claimants tell us a story, or the saint must have been a monster. They are also a little too apt to suppose every tombstone they dig up in Italy to have been a saint's or a martyr's; and they deify the names they find upon them, which for aught we know may have belonged to overseers of the road, or some of the greatest scoundrels in ancient Rome, or perhaps even to the persecutors of the primitive Christians.

SOLITUDE.

Hermits might have been very comfortable for aught I know, but I am persuaded there is no such thing, after all, as a perfect enjoyment of solitude; for, the more delicious the solitude, the more one wants a companion.

LOUIS XIV. AND GEORGE IV.

Louis XIV. was like George IV., inasmuch as he was fond of pleasure; but his ambition rendered him at once a better and a worse man than the latter, for it made him fonder of literature and the arts, which he knew would immortalize him, and it plunged him into a hundred useless wars, which the latter has never been able to undertake, and probably never would have undertaken, as he is so grossly indolent; for I do not think his virtues would preserve him from any error. In short, if the vices of Louis had greater opportunity to extend themselves than those of George, the Frenchman was, nevertheless, more sensible, more tasteful, more refined in his pleasures, more like a prince. He was more like the Emperor Augustus, except that he became a religious bigot in his old age—the common end of many a vicious man who is disappointed.

HENRY IV. OF FRANCE AND ALFRED.

My two favorite princes are Henry IV. of France and our own Alfred. The one, though he was a man of gallantry (which is to be pardoned, in a great measure, in a Frenchman of his time), was never depraved, never lost the goodness of his heart ; and he was a perfect hero of chivalry, as well as a philosopher in adversity. The other is the most perfect character in the list of monarchs of any age or country, a man who has come down to posterity without a single vice ; a warrior, a legislator, a poet, a musician, a philosopher ; a mixture of everything great and small that renders us dignified, wise, or accomplished ; a combination, indeed—

"Where ev'ry god did seem to set his seal
To give the world assurance of a man."

You see I must have recourse to Shakespeare. Nobody but such a writer can describe such a king.

FELLOWS OF COLLEGES.

These Fellows are absolute monks, without monkish superstition or restraint ; they live luxuriously, walk, ride, read, and have nothing to get, in this world, but a good appetite of a morning.

BEAUTY A JOY IN HEAVEN.

Beauty of every kind, *poeticized*, comes into the composition of my heaven—beauty of thinking, beauty of feeling, beauty of talking, beauty of hearing, and, of course, beauty of seeing, including visions of beautiful eyes and beautiful turns of limb.

ASSOCIATIONS OF GLASTONBURY.

Glastonbury is a town famous in old records for the most ancient abbey in the kingdom, for being the supposed birthplace of King Arthur, and for producing a species of whitethorn which was said to bud miraculously on Christmas-day; St. Joseph of Arimathea, it seems, having stuck his walking-stick in the ground on his arrival here, upon which the earth expressed its sense of the compliment by turning it into a thorn in blossom. Glastonbury is said to be the burial-place of King Arthur; but I am afraid the truth is, that he was buried in the same place in which he was born and lived—the brain of a poet.

LIBERTY OF SPEECH.

Whenever we feel ourselves in the possession of such a liberty and confidence of sociality as are not to be found in France or Turkey, then I must beg leave to return my thanks to the Hampdens, the Holts, Andrew Marvells, and other old English freemen, whose exertions, acting upon us to this very day, enable us to say and to enjoy what we do.

WRITING POETRY.

Poetry is very trying work, if your heart and spirits are in it—particularly with a weak body. The concentration of your faculties, and the necessity and ambition you feel to extract all the essential heat of your thoughts, seem to make up that powerful and exhausting effect called inspiration. The ability to sustain this, as well as all other exercises of the spirit, will evidently depend, in some measure, upon the state of your frame ; so that Dryden does not appear to have been altogether so fantastical in dieting himself for a task of verse ; nor Milton and others, in thinking their faculties stronger at particular periods : though the former, perhaps, might have rendered his caution unnecessary by undeviating temperance ; and the latter have referred to the sunshine of summer, or the in-door snugness of frosty weather, what they chose to attribute to a loftier influence.

THE WOMEN OF ITALY.

The *general* aspect of the women in Italy is striking, but not handsome ; that is to say, stronger marked and more decided than pleasing. But, when you do see fine faces, they are fine indeed; and they have all an intelligence and absence of affectation, very different from that idea of foreigners which the French are apt to give people.

FRENCH PEOPLE.

The French are pleasant in their manner, but seem to contain a good deal of ready-made heat and touchiness, in case the little commerce of flattery and sweetness is not properly carried on. There are a great many pretty girls, but I see no fine-faced old people, which is not a good symptom. Nor do the looks of the former contain much depth, or sentiment, or firmness of purpose. They seem made like their toys, not to last, but to play with and break up.

THE BLIND.

It is very piteous to look at blind people; but it is observed that they are generally cheerful because others pay them so much attention; and one would suffer a good deal to be continually treated with love.

LONDON.

London, as you say, is not a poetical place to look at; but surely it is poetical in the very amount and comprehensiveness of its enormous experience of pain and pleasure—a Shakespearean one. It is one of the great giant representatives of mankind, with a huge beating heart; and much of the vice even, and misery of it (in a deep philosophical consideration), is but one of the forms of the movement of a yet unsteadied progression, trying to balance things, and not without its reliefs; though, God knows, there is enough suffering to make us all keep a lookout in advance.

SOUTHEY'S POETRY.

I believe you are right about Southey's poetry, and cry mercy to it accordingly. He went to it too mechanically, and with too much nonchalance; and the consequence was a vast many words to little matter. Nor had he the least music in him at all. The consequence of which was, that he wrote prose out into lyrical wild shapes, and took the appearance of it for verse. Yet there was otherwise a poetical nature distributed through the mass, idly despising the concentration that would have been the salvation of it.

VULGAR CALUMNY.

I believe that one part of the public will always, if they can, calumniate any man who tries to amend them, and whom therefore they conclude to be their superior; but the great part, perhaps these included, will nevertheless be always willing to read one, provided they are amused by one's writings.

VALUE OF ACQUIREMENTS.

Acquirements of every sort increase our powers of doing good, both to ourselves and others; and the knowledge of languages—of *any* language almost—may turn out of the greatest service to us in advancing our prospects in life. The knowledge of French—and I have no doubt the case is the same with that of Italian, of Spanish, of German, etc.—has been known to give a young man great and sudden advantages over his fellows, and send him abroad upon the most interesting and important commissions. Suppose a messenger were required, for instance, to go on the sudden upon some urgent matter of government business to another country, and none were immediately to be had. A clerk starts up who understands Italian, and is dispatched in a hurry to Rome or Turin. Suppose an assistant botanist is required to explore an Eastern country; what an advantage the knowledge of Arabic or Persian would give him over competitors ignorant of those languages! Somebody has said that a man who understood four languages besides his own was five men instead of one.

THE BEARD.

Physicians proclaim it to be a "natural respirator"; it is manifestly a clothing and a comfort to the jaws and throat—ergo, probably, to regions adjoining; it is manly; it is noble; it is handsome. Think of all those beards of old, under tents and turbans; think of them now—how the whole East is bearded still, as it ever was, and ever will be, beard without end. The Chinese, it is true, are unbearded; but that was a Tartar doing, the work of the dynasty that is now being ousted. Confucius came before it, and had a beard as profound as his philosophy, you may rest assured. How else would the philosophy have come?—how have brooded to such purpose? —been so warm in his "nares" (as you justly observe) or so flowing toward his fellow creatures?

ATTRACTIONS OF HAM.

Old trees, the placidest of rivers, Thomson up above you, Pope near you, Cowley himself not a great way off: I hope here is a nest of repose, both material and spiritual, of the most Cowleian and Evelynian sort. Ham, too, you know, is expressly celebrated both by Thomson and Armstrong; and though that infernal Duke of Lauderdale lived there, who put people to the rack (in the *first* old original Ham House, I believe—

he married a Dysart), yet even the bitter taste of him is taken out of the mouth by the sweets of these poets, and by the memories of the good Duke and Duchess of Queensberry (Prior's Kitty), who nursed their friend Gay there when he was ill. Ay, and when he was well; and *upon* ham as well as in it; for you know he was a great eater, which made him, of course, ill again; and then they fed him on teas, and syllabubs, and ladies'-fingers, and again made him well, and able to be ill another time. And he was a punster too, was Gay, and doubtless punned as well as feasted on ham.

SLEEPING UNDER THE SKY.

The other day I had a delicious sleep in a haycock. These green fields and blue skies throw me into a kind of placid intoxication. Are there many moments more delicious than the one in which you feel yourself going to slumber, with the sense of green about you, of an air in your face, and of the great sky arching over your head? One feels, at such times, all the grandeur of planetary consciousness without the pain of it. You know what I mean. There is a sort of kind and beautiful sensuality in it which softens the cuts and oppressiveness of intellectual perception. Certainly, a country so green as England can not well be equaled by any other at such a season.

WAR POETRY.

You may judge what I felt about the war sonnets, when I opened the book on the one beginning—

"Blaze gun to gun," etc., etc.,

with that affectation of encouraging "living mires," and "hells of fire," which is or ought to be revolting to a poet's heart, and is not at all his business : for to say it is necessary to oppose the "commonplaces of humanity" with such outrages upon them, is itself a commonplace, however it may seem otherwise to the unreflecting. Mankind are always too ready to continue the barbarism, war ; and whatever may be the unavoidableness of it, or even the desirableness of it, at some particular moment, when forced upon us by barbarism itself, it is not the poet's business to lay down his harp of Orpheus and halloo brutalities on.

And as to God's permission, and therefore use of such things, we might as well encourage, instead of piously helping to do away, any other evils through which, or in spite of which, good mysteriously progresses, and strike up howls in praise of murder in ordinary and Bartholomew massacres. Such mistakes vex one in men of genius, who ought to know better.

MONEY-GETTING.

You are right about money-getting in the main, horrible as are the abuses of it, and provoking sometimes its predominance. Besides, it is a phase of things through which all the world must go, till they have all made acquaintance with one another, and all interchanged their goods and knowledges; by which time it is to be hoped they will all have discovered the means and advantages of obtaining more leisure, varying the pursuit, and exalting its objects: for I suppose we are not to believe that the world is to go on through countless millions of ages precisely as it does at this or any other moment, merely because Jones trades with Thompson, and Smith is a pork-butcher.

VALUE OF WORDS.

Words are often things also, and very precious, especially on the gravest occasions. Without "words," and the truth of things that is in them, what were we?

UNWRITTEN REVELATIONS.

The only two books of paramount authority with me are the Book of Nature, and the heart of its reader, Man; and that the operations in the one, and the aspirations of the other (though I fully concede, as I am bound to do, all the reconcilements, and possibilities, and transcendentations of every kind, which greater understandings and imaginations than my own may see in other books), compel me—if so glad a conclusion can be called compulsion—to be of the opinion that God is the unmingled, wholly benevolent, and conscious spirit of Good, working through His agent, Man; that evil, where it *is* evil, and not a necessary portion of good (as it probably all is ultimately), is the difficulty presented to the course of this working by the unconscious, involuntary, and therefore unmalignant mystery called Matter; that God, though not immediately or in all stages of His processes almighty, is ultimately so; and that His constant occupation is the working out of heavens in place and time, in which prospection and retrospection somehow or other become reconciled to the final conscious beatitude of all the souls that have ever existed.

WEEPING.

It is an affecting, and would be a startling consideration, to think that God has given us tears for such express purposes of relief, as knowing how much our sorrows would need them, were not this very fact, among others, a proof (at least, it is a great evidence to myself) that all other needs of our affections are destined to be made up to us in good time. For tears, though they calm the first outbreaks of affliction, do not suffice for its subsequent yearnings; and as those yearnings continue—often with great returns of anguish to the last—sufficingness, I think, remains in store for them also. I should be one of the unhappiest, instead of the most resigned of men, at this moment, if I did not constantly, and as it were instinctively, feel that I should rejoin all the dear ones whom I have lost—words that now, as I write, wring bitter and unsufficing tears from the quivering of the soul within me. Encourage and, as it were, throw yourself heartily into the arms of this expectation; think how worthy it is, both of man and God, quite apart from the dogmas which too often render both so much the reverse; and, meantime, act in every respect with regard to your dear one just as you feel sure *she would wish you to act*, weeping as plentifully as you need, but as patiently too, and considering her as only gone before you, to be rejoined: she, all the while, being delivered from all *her* pain, spiritual as well as bodily, because she now possesses that certainty, as a disembodied spirit, which, for some finally good purpose, it is not fit that we, who are yet on earth, should possess ourselves. For my part, I confess to you that I often feel it highly probable that the spirits of my own beloved dead are in the room with me, and that they feel a special and heavenly pleasure by seeing that I do so, and by knowing the comfort it gives me.

I count this no kind of madness, but one of the heights of reason ; for it does not unfit me for the common work of life, but, on the contrary, helps it. And as it neither fevers me, nor is caused by any fever itself, I count it not among the unhealthy, but the healthy capabilities of my nature ; therefore of anybody else's nature who chooses reasonably to enjoy it.

IMAGINARY CONVERSATIONS

OF

POPE AND SWIFT.

CONVERSATION OF POPE.

REPORTED BY A YOUNG GENTLEMAN WHO DINED WITH HIM.

July 4, 1727.

YESTERDAY was a day of delight. I dined with Mr. Pope. The only persons present were the venerable lady his mother, Mrs. Martha Blount, and Mr. Walscott, a great Tory, but as great a lover of Dryden; which Mr. Pope was pleased to inform me was the reason he had invited me to meet him. Mr. Pope was in black, with a tie-wig. I could not help regarding him, as he sat leaning in his arm-chair before dinner, in the light of a portrait for posterity. When he came into the room, after kindly making me welcome, he took some flowers out of a little basket that he had brought with him, and presented them, not to Mrs. Martha, who seemed to look as if she expected it, but to Mrs. Pope; which I thought very pretty, and like a gentleman, not in the ordinary way. But the other had no reason to be displeased; for, turning to her with the remainder, he said: "I was thinking of a compliment to pay you; so I have done it." He flatters with as much delicacy as Sir Richard Steele; and the ladies like it as much from him. What fine-shaped fellows have I seen, who could not call up half such looks into their eyes!

I was in a flutter of spirits, which took away my appetite. Mr. Pope recommended his fish and his Banstead mutton to no purpose. I was too well fed with hearing him talk. However, I mechanically drank his wine, which emboldened me to say something. What I said, I do not very well remember, and it is no matter. I have even forgotten some agreeable stories related by Mr. Walscott about the civil wars; but every word that passed the lips of Mr. Pope seems engraven on my brain. From the subject of killing mutton, the talk fell upon cruelty to animals, upon which Mr. Pope made some excellent observations. He began by remarking how strange it was that little or nothing had been said of it in books.

Mr. Walscott. I suppose authors have been too much in the habit of attending to the operation of their own minds.

Mr. Pope. But they have been anglers. I

have a curious book in my library written by one Isaac Walton, an old linen-draper in the time of Charles the Second, who was fond of meadows and village ale-houses, and has really a pretty pastoral taste. This man piques himself on his humanity; and yet the directions he gives on the subject of angling (for the book is written on that art) are full of such shocking cruelty that I do not care to repeat them before ladies. He wrote the lives of Donne, Hooker, and others, all anglers, and good religious men. Yet I suppose they were all as cruel. It is wonderful how the old man passes from pious reflections to the tortures of fish and worms, just as if pain were nothing. Yet what else are the devil and his doings made of?

MR. WALSCOTT. Dryden was an angler.

MR. POPE. Yes; he once exclaimed of D'Urfey, "*He* fish!" because the man attempted to write. There is a passage in his "Astræa Redux," written in the proper fishing spirit; that is to say, in which all the consideration is for the fisher, and none for the fish.

MR. WALSCOTT. I remember it. He is speaking of General Monk, and the way in which he brought about the grand stroke for the Restoration:

> " 'Twas not the hasty product of a day,
> But the well-ripened fruit of wise delay.
> He, like a patient angler, ere he strook,
> Would let him play a while upon the hook."

Mr. Pope. The "patient angler"! Mighty patient truly, to sit at a man's ease and amuse himself! The question is, what the fish think of it.

Mrs. Martha Blount. Sure it must be so; and yet I never once thought of that before. God forgive me for the murders I committed last year in Oxfordshire, at the instigation of my brother!

Mr. Pope looked at her with benevolence as she said this; but he was too much in earnest to pay her the compliments which ordinary gallantry would have struck out of the confession. I really believe he feels as much for carp and trout as most men do for each other.

Mr. Walscott. But would it not be exchanging one pain for another, to make people think too much of these things?

Mr. Pope. That is well said. But I know not what right we have to continue putting our fellow creatures to pain, for the sake of avoiding it ourselves. Besides, there is a pain that exalts the understanding and morals, and is not unallied with pleasure; which can not be said of putting hooks into poor creatures' jaws and bowels.

Mr. Walscott. There is a good deal in that. Yet all animals prey upon one another. We prey upon them ourselves. We are at this minute availing ourselves of the cruelties of butchers and fishermen.

Mr. Pope. Not the cruelties. Killing and torturing are different. Death is inevitable to all; and a sheep who has passed his days in the meadows, and undergone a short death from a knife, has had as good a bargain as most of us. Animals kill, but they do not torture one another.

Mr. WALSCOTT. I think I have read of instances. Yes, I am sure of it; and what think you of the cat with a mouse?

Mr. POPE. Why, I think she is very like an angler. I should wish to see a treatise on the subject by a cat. It is a meditative creature, like old Isaac, and is fond of fish. I am glad to see how much the *fera natura* excuses them both; but to us, who can push our meditations further, the excuse is not the same.

Mr. WALSCOTT. Yet this appears to be instinct. What say you to Nature? It is her own doing.

Mr. POPE. Nature is a very wide term. We make use of it rather to get rid of arguments than to enforce them. If it is the cat's nature to torment, it is man's nature to know better. Improvement is nature. The reflections we are now making are nature. I was wrong in saying that no animal tortures another; but pray observe— we abuse animals, when it suits us, as the brute creation, and call upon them to bear testimony to our natural conduct, when we are pleased to resemble them. Now, the matter is, that we ought to imitate them solely in what is good and beneficial; and in all other cases, give both them and ourselves the benefit of our better knowledge.

Mr. WALSCOTT. Evil will exist in spite of us.

Mr. POPE. I do not know that. It is impossible for us, who only see to the length of a little miserable point in the midst of eternity, to say what will or will not exist. But we must give our fellow creatures the benefit of our knowledge, and our ignorance, too. If we can not abolish evil, we may diminish it, or divide it better; and nature incites us to do so by putting the thought in our heads. It is fancied by some, and I dare say anglers fancy it, that animals, different from us in their organization, do not feel as we do. I hope not. It is at least a good argument for consolation, when we can do nothing to help them; but, as we are not sure of it, it is an argument not to be acted upon when we can. They must have the benefit of our want of certainty. Come, anglers shall have the benefit of it, too. Old Walton was as good a man as you could make out of an otter; and I like the otter the better for him. Dryden, I am sure, was humane: he was too great a man to be otherwise. But he had all his bodily faculties in perfection; and I sometimes think that animal spirits take the place of reflection on certain animal occasions, and fairly occupy the whole man instead of it, even while he thinks he's thinking. Yet I am afraid Donne and the others sophisticated; for subtlety was their business. There are certain doctrines that do men no good, when the importance of a greater or less degree of pain in this world comes to be made a question of; and so they get their excuse that way. Anything rather than malignity and the determination to give pain; and yet I know not how the angler is to be found guiltless on that score, if he reflects on what he is going about. I am sure he must hurt his own mind, and perplex his ideas of right and wrong.

Mr. Walscott concluded the argument by owning himself much struck with the variety of reflections which Mr. Pope had brought forward or suggested. He said he thought they would make a good poem. Mr. Pope thought so too, if enlivened with wit and description; and said he should, perhaps, turn it in his mind. He remarked that, till the mention of it by Sir Richard Steele, in the "Tatler," he really was not aware that anything had been said against cruelty to animals by an English writer, with the exception of the fine hint in Shakespeare about the beetle. "Steele," said he, "was then a gay fellow about town, and a soldier; yet he did not think it an imputation on his manhood to say a good word for tomtits and robins. Shakespeare, they tell us, had been a rural sportsman; and yet he grew to sympathize with an insect." I mentioned the "Rural Sports" of Mr. Gay, as enlisting that poet among the anglers that rejected worms. "Yes," said he, "Gay is the prettiest *fera natura* that ever was, and catches his trout handsomely to dine upon. But you see the effect of habit, even upon him. He must lacerate fish, and yet would not hurt a fly. Dr. Swift, who loves him as much as he hates angling, said to him one day at my Lord Bolingbroke's, 'Mr. Gay, you are the only angler I ever heard of, with an idea in his head; and it is the only idea you have, not worth having.' Angling makes the Dean melancholy, and sets him upon his yahoos."

This authority seemed to make a greater impression upon Mr. Walscott than all the reasoning. He is a very great Tory, and prodigiously admires the Dean. Mr. Pope delighted him by asking him to come and dine with them both next week; for the Dean is in England, and Mr. Pope's visitor. I am to be there too. "But," says he, "you must not talk too much about Dryden; for the Doctor does not love him." Mr. Walscott said he was aware of that circumstance from the Dean's works, and thought it the only blemish in his character. For my part, I had heard a story of Dryden telling him he would never be a poet; but I said nothing. Mr. Pope attributed his dislike to a general indignation he felt against his relations, for their neglect of him when young. For Dryden was his kinsman. The Davenants are his relations, and he does not like them. Mr. Walscott asked if he was an Englishman or an Irishman; for he never could find out. "You would find out," answered Pope, "if you heard him talk; for he can not get rid of the habit of saying *a* for *e*. He would be an Englishman with all his heart if he could; but he is an Irishman, that is certain, and with all his heart too, in one sense; for he is the truest patriot that country ever saw. He has the merit of doing Ireland the most wonderful services, without loving her; and so he does to human nature, which he loves as little; or at least he thinks so. This, and his wit, is the reason why his friends are so fond of him. You must not talk to him about Irish rhymes," added Mr. Pope, "any more than you must talk to me about the *gods* and *abodes* in my Homer, which he quarrels with me for. The truth is, we all write Irish rhymes; and the Dean contrives to be more exact in that way than most of us." "What!" said Mr. Walscott, "does he carry his Irish accent into his writings, and yet think to conceal himself?" Mr. Pope read to us an odd kind of Latin-English effusion of the Dean's, which made us shake with laughter. It was about a consultation of physicians. The words, though Latin themselves, make English when put together; and the Hibernianism of the spelling is very plain. I remember a taste of it. A doctor begins by inquiring:

"Is his Honor sic? Præ lætus felis pulse. It do es beat veris loto *de*."*

Here *de* spells *day*. An Englishman would have used the word *da*.

"No," says the second doctor, "no notis as qui cassi e ver fel tu metri it," † etc., etc.

Metri for *may try*.

Mr. Pope told us that there were two bad rhymes in the "Rape of the Lock," and in the space of eight lines—*side* and *subside,* and *endued* and *subdued*.‡

Mr. Walscott. Those would be very good French rhymes.

Mr. Pope. Yes, the French make a merit of necessity, and force their poverty upon us for riches. But it is bad in English. However, it is too late to alter what I wrote. I now care less about them, notwithstanding the Doctor. When I was a young man, I was for the free disengaged way of Dryden, as in the "Essay on Criticism"; but the town preferred the style of my "Pastorals," and somehow or other I agreed with them. I then became very cautious, and wondered how those rhymes in the "Lock" escaped me. But I have now come to this conclusion: that when a man has established his reputation for being able to do

---

* Is his Honor sick? Pray let us feel his pulse. It does beat very slow to-day.

† No, no, no! 'tis as quick as I ever felt. You may try it.

‡ Vide pp. 120, 121 of the present volume.

a thing, he may take liberties. Weakness is one thing, and the carelessness of power another. This makes all the difference between those shambling ballads that are sold among the common people and the imitations of them by the wits to serve a purpose; between Sternhold and Hopkins, and the ballads on the Mohocks and great men.

Mr. Pope then repeated, with great pleasantry, Mr. Gay's verses in the "Wonderful Prophecy":

"From Mohock and from Hawkubite,
    Good Lord, deliver me!
Who wander through the streets by night
    Committing cruelty." *

Mr. Walscott, with all his admiration of Dryden, is, I can see, a still greater admirer of the style of Mr. Pope. But his politics hardly make him know which to prefer. I ventured to say that the "Rape of the Lock" appeared to me per-

* The other verses, which Mr. Pope's visitor has not set down, are as follows:

"They slash our sons with bloody knives,
    And on our daughters fall;
And if they ravish not our wives,
    We have good luck withal.
Coaches and chairs they overturn,
    Nay, carts most easily;
Therefore from Gog and eke Magog,
    Good Lord, deliver me!"

The Mohocks were young rakes, of whom terrible stories were told. They were said to be all of the Whig party.

fection; but that still, in some kinds of poetry, I thought the licenses taken by the "Essay on Criticism" very happy in their effect: as, for instance, said I, those long words at the end of couplets:

> "Thus, when we view some well-proportioned dome
> (The world's just wonder, and e'en thine, O Rome!)
> No single parts unequally surprise;
> All come united to the admiring eyes;
> No monstrous height, or breadth, or length appear;
> The whole at once is bold and regular."

Now here, I said, is the regularity and the boldness too. And again:

> "'Twere well might critics still this freedom take;
> But Appius reddens at each word you speak,
> And stares tremendous with a threatening eye,
> Like some fierce tyrant in old tapestry."

And that other couplet:

> "With him most authors steal their works, or buy;
> Garth did not write his own *Dispensary*."

I said, this last line, beginning with that strong monosyllable, and throwing off in a sprightly manner the long word at the end, was like a fine bar of music, played by some master of the violin. Mr. Pope smiled, and complimented me on the delicacy of my ear, asking me if I understood music. I said no, but was very fond of it. He fell into a little musing, and then observed that he did not know how it was, but writers fond of music appeared to have a greater indulgence for the licenses of versification than any others. The two smoothest living poets were not much attached to that art. (I guess he meant himself and Dr. Swift.) He inquired if I loved painting. I told him so much so that I dabbled in it a little myself, and liked nothing so much in the world, after poetry. "Why, then," said he, "you and I, some fine morning, will dabble in it like ducks." I was delighted at the prospect of this honor, but said I hoped his painting was nothing nigh equal to his poetry, or I would not venture to touch his palette. "Oh," cried he, "I will give you confidence." He rose with the greatest good nature, and brought us a sketch of a head after Jervis, and another of Mrs. Martha. I had begun to fear that they might be unworthy of so great a man, even as amusements; but they were really wonderfully well done. I do think he would have made a fine artist, had he not been a poet.* He observed that we wanted good criticism on pictures; and that the best we had yet were some remarks of Steele's in the "Spectator," on the cartoons of Raphael. He added a curious observation on Milton: that with all his regard for the poets of Italy, and his travels in that country, he has said not a word of their painters, nor scarcely alluded to painting throughout his works.

* This has been doubted by others who have seen his performances. Some of them remain, and are not esteemed.

Mr. Walscott. Perhaps there was something of the Puritan in that. Courts, in Milton's time, had a taste for pictures: King Charles had a fine taste.

Mr. Pope. True; but Milton never gave up his love of music—his playing on the organ. If he had loved painting, he would not have held his tongue about it. I have heard somebody remark that the names of his two great archangels are those of the two great Italian painters, and that their characters correspond; which is true and odd enough. But he had no design in it. He would not have confined his praises of Raphael and Michael Angelo to that obscure intimation. I believe he had no eyes for pictures.

Mr. Walscott. Dryden has said fine things about pictures. There is the epistle to your friend Sir Godfrey, and the ode on young Mrs. Killigrew. Did he know anything of the art?

Mr. Pope. Why, I believe not; but he dashed at it in his high way, as he did at politics and divinity, and came off with flying colors. Dryden's poetic faith was a good deal like his religious. He could turn it to one point after another, and be just enough in earnest to make his belief be taken for knowledge.

Mr. Pope told us that he had been taken, when a boy, to see Dryden at a coffee-house. I felt my color change at this anecdote, so vain do I find myself. I took the liberty of asking him how he felt at the sight; for it seems he only saw Dryden; he did not speak to him, which is a pity.

MR. POPE. Why, I said to myself: "That is the great Mr. Dryden; there he is: he must be a *happy man.*" This notion of his happiness was the uppermost thing in my mind, beyond even his fame. I thought a good deal of that; but I knew no pleasure, even at that early age, like writing verses; and there, said I, is the man who can write verses from morning till night, and the finest verses in the world. I am pretty much of the same way of thinking now. Yes; I really do think that I could do nothing but write verses all day long, just taking my dinner, and a walk or so, if I had health. And I suspect it is the same with all poets—I mean with all who have a real passion for their art. Mr. Honeycomb, I know, agrees with me, from his own experience.

The gratitude I felt for this allusion to what I said to him one day at Button's was more than I can express. I could have kissed his hand out of love and reverence. "Sir," said I, "you may guess what I think of the happiness of poets, when it puts me in a state of delight inconceivable to be supposed worthy of such a reference to my opinion." I was indeed in a confusion of pleasure. Mr. Walscott said it was fortunate the ladies had left us, or they might not have approved of such a total absorption in poetry. "Oh!" cried Mr. Pope, "there we have you; for the ladies are a part of poetry! We do not leave them out in our studies, depend upon it."

I asked him whom he looked upon as the best love-poet among our former writers. I added "former," because the "Epistle of Héloïse to Abelard" appears to me to surpass any express poem on that subject in the language. He said Waller; but added, it was after a mode. "Everything," said Mr. Pope, "was after a mode, then. The best love-making is in Shakespeare. Love is a business by itself in Shakespeare, just as it is in nature."

MR. WALSCOTT. Do you think Juliet is natural when she talks of cutting Romeo into "little stars," and making the heavens fine?

MR. POPE. Yes; I could have thought that, or anything else, of my mistress, when I was as young as Romeo and Juliet. Petrarch, as somebody was observing the other day, is natural for the same reason, in spite of the conceits which he mixes up with his passion; nay, he is the more natural, supposing his passion to have been what I take it—that is to say, as deep and as wonder-working as a boy's. The best of us have been spoiled in these matters by the last age. Even Mr. Walsh, for all his good sense, was out in that affair, in his Preface. He saw very well that a man, to speak like a lover, should speak as he felt; but he did not know that there were lovers who felt like Petrarch.

Mr. Walscott. You would admire the writings of one Drummond, a Scotch gentleman, who was a great loyalist.

Mr. Pope. I know him well, and thank you for reminding me of him. If he had written a little later here in England, and been published under more favorable circumstances, he might have left Waller in a second rank. He was more in earnest, and knew all points of the passion. There is great tenderness in Drummond. He could look at the moon, and think of his mistress, without thinking how genteelly he should express it, which is what the other could not do. No; we have really no love-poets, except the old dramatic writers; nor the French, either, since the time of Marot.

Mr. Walscott. And very pretty writing it is, if managed as Mr. Pope manages it.

Mr. Pope. I do not undervalue it, I assure you. After Shakespeare, I can still read Voiture, and like him very much : only it is like coming from country to town, from tragedy to the ridotto. To tell you the truth, I am as fond of the better sort of those polite writers as any man can be ; and I feel my own strength to lie that way ; but I pique myself on having something in me besides, which they have not   I am sure I should not have been able to write the "Epistle of Héloïse," if I hadn't. There is a force and sincerity in the graver love-poets, even on the least spiritual parts of the passion, which writers, the most ostentatious on that score, might envy.

Mr. Walscott. The tragedy of love includes the comedy, eh?

Mr. Pope. Why, that is just about the truth of it, and is very well said.

Mr. Pope's table is served with neatness and elegance. He drinks but sparingly. His eating is more with an appetite, but all nicely. After dinner, he set upon table some wine given him by my Lord Peterborough, which was excellent. He then showed us his grotto, till the ladies sent to say tea was ready. I never see a tea-table but I think of the "Rape of the Lock." Judge what I felt when I saw a Mrs. Fermor, kinswoman of Belinda, seated next Miss Martha Blount, who was making tea and coffee. There was an old lady with her; and several neighbors came in from the village. This multitude disappointed me, for the talk became too general; and my lord's wine, mixed with the other wine and the wit, having got a little in my head, and Mr. Pope's attention being repeatedly called to other persons, I can not venture to put down any more of his conversation. But I shall hear him again, and, I hope, again and again. So patience till next week.

CONVERSATION OF SWIFT AND POPE.

RECORDED BY THE SAME VISITOR.

July 15, 1727.

AT length the *dies optanda* came. Shall I confess my weakness? I could do nothing all the morning but walk about, now reading something of the Dean's or Mr. Pope's, and now trying to think of some smart things to say at dinner! I did not say one of them. Yes, I made an observation on Sannazarius, which was well received. I must not forget the boatman who took me across the water from Sutton. "Young gentleman," says he, "if I may make so bold, I will tell you a piece of my mind." "Well, pray do." "Why, I'm thinking you're going to see your sweetheart, or else the great poet yonder, Mr. Pope." "Why so?" said I, laughing. "Why," said he, "your eyes are all in a sparkle, and you seem in a woundy hurry." I told him he had guessed it. He is in the habit of taking visitors over; great lords, he said, and grand ladies from court; "and very merry, too, for all that." He mentioned Dr. Swift, Mr. Gay, and others. Upon asking if Dr. Swift was not one of the great writers, "Ay, ay," said he, "let him alone, I warrant him: he's a strange gentleman." The boatman told me that one day the Dean, "as they called him," quarreled with him about a halfpenny. His Reverence made him tack about for some whimsey or other, and then would not pay him his due, because he did not tell him what the fare was the moment he asked. "So his Deanship left a cloak in the boat, and I took it up to him to Mr. Pope's house, and he came out and said, 'Well, sirrah, there's some use in frightening you sneaking rascals, for you bring us back our goods.' So I thought it very strange; and says I, 'Your Reverence thinks I was frightened, eh?' 'Yes,' says he, as sharp as a needle; 'haven't you done an honest action?' So I was thrown all of a heap to hear him talk in such a way; and as I didn't well know what he meant, I grew redder and redder like, for want of gift of the gab. So says I at last, 'Well, if your Reverence, or Deanship, or what you please to be called, thinks as how I was frightened, all that I says is this: d—n me (saving your Reverence's presence) if Tom Harden is a man to be frightened about a halfpenny, like some folks that shall be nameless.' 'Oh, ho!' says Mr. Dean, looking scared, like an owl in an ivy-bush, 'Tom Harden is a mighty pretty fellow, and must not be flouted; and so he won't row me again, I suppose, for all he has got a wife and a parcel of brats.' How he came to know that, I can't say. 'No, no,' says I; 'I'm not so much of a pretty fellow as that comes to, if that's what they mean by a pretty fellow. It's not my business to be picking and choosing my fares, so that I gets my due. But I was right about the halfpenny for all that; and if your Reverence wants to go a swan-hopping another time, you knows what's to pay.' So the Dean fell a-laughing like mad, and then looked very grave, and said, 'Here, you, Mr. John Searle' (for that's Mr. Pope's man's name)—'here, make Mr. Thomas Harden acquainted with the taste of your beer; and do you, Mr. Thomas, take back the cloak, and let it stay another time in the boat till I want to return; and, moreover, Thomas, keep the cloak always for me to go home o' nights in, and I will make it worth your while, and leave it you when I am dead, provided it's worn out enough.' I shall never forget all the odd things he said, for I talked 'em over with Mr. Searle. 'And hark'ee, Mr. Thomas Harden,' says he, 're- member,' says he, 'and never forget it, that you love your wife and children better than your pride,' says he, 'and your pride,' says he, 'better than a paltry Dean; and those are two nice things to manage together.' And the Dean has been as good as his word, young gentleman; and I keep his cloak; and he came to my cot- tage yonder one day, and told my wife she was 'the prettiest creature of a plain woman' that he ever saw (did you ever hear the like o' that?); and he calls her Pannopy, and always asks how she does. I don't know why he calls her Pannopy — mayhap because her pots and pans were so bright; for you'd ha' thought they'd been silver, from the way he stared at 'em." *

Having heard of the Dean's punctuality, I was afraid I should be too late for my good behavior; but Mr. Thomas reassured me by saying that he had carried his Reverence across three hours before from Richmond, with Madam Blount. "He is in a mighty good humor," said he, "and will make you believe anything he likes, if you don't have a care."

I was in very good time, but found the whole party assembled with the exception of Mrs. Pope. It was the same as before, with the addition of the Doctor. He is shorter and stouter than I had fancied him, with a face in which there is nothing remarkable, at first sight, but the blueness of the eyes. The boatman, however, had not prepared me for the extreme easiness and good-breeding of his manners. I had made a shallow conclusion. I expected something perpetually fluctuating between broad mirth and a repelling self-assumption. Nothing could be more unlike what I found. His mirth, afterward, was at times broad enough, and the ardor and freedom of his spirit very evident; but he has an exquisite mode, throughout, of

---

* Probably from a strange line in Spenser, where he describes the bower of Proteus:

"There was his wonne; no living wight was scene,
Save one old nymph, hight Panope, to keep it cleane."
—*Faerie Queen*, book iii.

maintaining the respect of his hearers. Whether
he is so always, I can not say. But I guess that
he can make himself equally beloved where he
pleases, and feared where he does not. It must
be owned that his mimicry (for he does not disdain even that sometimes) would not be so well
in the presence of foolish people. I suppose he
is cautious of treating them with it. Upon the
whole, partly owing to his manners, and partly
to Mr. Pope's previous encouragement of me
(which is sufficient to set up a man for anything),
I felt a great deal more at my ease than I expected, and was prepared for a day as good as the
last. One of the great arts, I perceive, of these
wits, if it be not rather to be called one of the
best tendencies of their nature (I am loath to bring
my modesty into question by saying what I think
of it), is to set you at your ease, and enlist your
self-love in their favor, by some exquisite recognition of the qualities or endeavors on which you
most pride yourself, or are supposed to possess.
It is in vain you tell yourself they may flatter you.
You believe and love the flattery; and let me add
(though at the hazard of making my readers smile),
you are bound to believe it, if the bestowers are
men of known honesty and spirit, and above "buying golden opinions" of everybody. I am not sincere when I call it an art. I believe it to be good-
natured instinct, and the most graceful sympathy;
and having let this confession out, in spite of my-

self, I beg my dear friends, the readers, to think the best they can of me, and proceed. The Dean is celebrated for a way he has of setting off his favors in this way, by an air of objection. Perhaps there is a little love of power and authority in this, but he turns it all to grace. Mr. Pope did me the honor of introducing me as a young gentleman for whom he had a particular esteem. The Dean acknowledged my bow in the politest manner; and after asking whether this was not the Mr. Honeycomb of whom he had heard talk at the coffee-house, looked at me with a serious calmness, and said, "I would not have you believe, sir, everything Mr. Pope says of you." I believe I blushed, but without petulance. I answered that my self-love was doubtless as great as that of most young men, perhaps greater; and that if I confessed I gave way to it in such an instance as the present, something was to be pardoned to me on the score of the temptation. "But," said he, "Mr. Pope flatters beyond all bounds. He introduces a new friend to us, and pretends that we are too liberal to be jealous. He trumpets up some young wit, Mr. Honeycomb, and fancies, in the teeth of all evidence, moral and political, that we are to be in love with our successors." I bowed and blushed, indeed, at this. I said that, whether a real successor or not, I should now, at all events, run the common danger of greatness, in being spoilt by vanity; and that, like a subtle prince in possession, the Dean knew how to prevent his heirs presumptive from becoming of any value. The Doctor laughed, and said, with the most natural air in the world: "I have read some pretty things of yours, Mr. Honeycomb, and am happy to make your acquaintance. I hope the times will grow smoother as you get older, and that you will furnish a new link, some day or other, to reunite friends that ought not to have been separated." This was an allusion to certain Whig patrons of mine. It affected me much; and I gladly took the opportunity of the silence required by good-breeding, to lay my hand upon my heart, and express my gratitude by another bow. He saw how nearly he had touched me; for, turning to Mr. Pope, he said gayly: "There is more love in our hates nowadays than there used to be in the loves of the wits, when you and I were as young as Mr. Honeycomb. What did you care for old Wycherley? or what did Wycherley care for Rochester, compared with the fond heats and vexations of us party-men?" Mr. Pope's answer was prevented by the entrance of his mother. The Dean approached her as if she had been a princess. The good old lady, however, looked as if she was to be upon her good behavior, now that the Dean was present; and Mrs. Martha Blount, notwithstanding he pays court to her, had an air of the same kind. I am told that he keeps all the women in awe. This must be one of the reasons for their being so fond of him, when he chooses to be pleased. Mr. Walscott, whose manners are simple and sturdy, could not conceal a certain uneasiness of admiration; and, though a great deal more at my ease than I had looked to be, I partook of the same feeling. With Mr. Pope, all is kindness on one part, and pleased homage on the other. Dr. Swift keeps one upon the alert, like a field-officer. Yet, externally, he is as gentle, for the most part, as his great friend.

The dinner seemed to be still more neat and perfect than the last, though I believe there were no more dishes. But the cookery had a more consummate propriety. The Dean's influence, I suppose, pierces into the kitchen. I could not help fancying that the dishes were sensible of it, and submitted their respective relishes with anxiety. The talk, as usual, began upon eating.

MR. POPE. I verily believe, that when people eat and drink too much, if it is not in the ardor of good company, they do it, not so much for the sake of eating, as for the want of something better to do.

DR. SWIFT. That is as true a thing as you ever said. When I was very solitary in Ireland, I used to eat and drink twice as much as at any other time. Dinner was a great relief. It cut the day in two.

MR. POPE. I have often noticed, that if I am

alone, and take up a book at dinner-time, and get concerned in it, I do not care to eat any more. What I took for an unsatisfied hunger leaves me —is no more thought of.

DR. SWIFT. People mean as much when they say that such and such a thing is meat and drink to them. By the same rule, meat and drink is one's book. At Laracor, an omelet was Quintus Curtius to me; and the beef, being an epic dish, Mr. Pope's Homer.

MR. WALSCOTT. You should have dressed it yourself, Mr. Dean, to have made it as epic as that.

DR. SWIFT. 'Faith! I was no hero, and could not afford the condescension. A poor vicar must have a servant to comfort his pride, and keep him in heart and starvation.

MR. WALSCOTT. If people eat and drink for want of something better to do, there is no fear that men of genius will die of surfeiting. They must have their thoughts to amuse them, if nothing else.

THE DEAN (with vivacity). Their thoughts! Their fingers' ends, to bite till the blood come. That, Mr. Walscott, depends on the state of the health. I was once returning to dinner at Laracor, when I saw a grave little shabby-looking fellow sitting on a stile. I asked him what he did idling there. He answered, very philosophically, that he was the Merry Andrew lately arrived, and that, with my leave, he would drink my health in a little more fresh air, for want of a better draught. I told him I was a sort of Merry Andrew myself, and so invited him to dinner. The poor man became very humble and thankful, and turned out a mighty sensible fellow; so I got him a place with an undertaker, and he is now merry in good earnest. I put some pretty "thoughts" in his head before he left me. A cousin of mine sent them me from Lisbon, in certain long-necked bottles, corked and sealed up. My Lord Peterborough has a cellar full of very pretty thoughts. God grant we all keep our health! and then, young gentleman (looking very seriously at me, for I believe he thought my countenance expressed a little surprise)—and then we shall turn our thoughts to advantage for ourselves and for others.

Mrs. Pope. If there's any gentleman who could do without his wine, I think it must be my lord. When I was a little girl, I fancied that great generals were all tall stately persons, with one arm akimbo, and a truncheon held out in the other; and I thought they all spoke grand, and like a book.

Dr. Swift. Madam, that was Mr. Pope's poetry, struggling to be born before its time.

Mrs. Pope. I protest, when I first had the honor of knowing my Lord Peterborough, he almost frightened me with his spirits. I believe he saw it; for all of a sudden he became the finest, softest-spoken gentleman that I ever met with; and I fell in love with him.

Mrs. BLOUNT. Oh, madam, I shall tell! and we'll all dance at my lady's wedding.

I do not know which was the handsomer sight; the little blush that came over the good old lady's cheek as she ended her speech, or the affectionate pleasantness with which her son regarded her.

MR. POPE. You did not fall in love with Lord Peterborough because he is such a fine-spoken gentleman, but because he is a fine gentleman and a madcap besides. I know the tastes of you ladies of the civil wars.

THE DEAN. 'Tis a delicious rogue! (and then, as if he had spoken too freely before strangers)— 'tis a great and rare spirit! If all the world resembled Lord Peterborough, they might do without consciences. I know no fault in him, but that he is too fond of fiddlers and singers.

MR. POPE. Here is Mr. Honeycomb, who will venture to dispute with you on that point.

I said Mr. Pope paid me too great a compliment. I might venture to differ from Dr. Swift, but hardly to dispute with him.

DR. SWIFT. Oh, Mr. Honeycomb, you are too modest, and I must pull down your pride. You have heard of little Will Harrison, poor lad, who wrote the "Medicine for the Ladies," in the "Tatler." Well, he promised to be one of your great wits, and was very much of a gentleman; and so he took to wearing thin waistcoats, and died of a birthday suit. Now, thin waistcoats and soft sounds are both of 'em bad habits, and encourage a young man to keep late hours, and get his death o' cold.

I asked whether he could not admit a little "higher argument" in the musician than the tailor. Shakespeare says of a flute, that it "discoursed excellent music," as if it had almost been a rational creature.

Dr. Swift. A rational fiddlestick! It is not Shakespeare that says it, but Hamlet, who was out of his wits. Yes, I have heard a flute discourse. Let me see—I have heard a whole roomful of 'em discourse (and then he played off an admirable piece of mimicry, which ought to have been witnessed, to do it justice). Let me see— let me see. The flute made the following excellent remarks: *Tootle, tootle, tootle, tootle—tootle, tootle tee;* and then again, which I thought a new observation, *Tootle, tootle, tootle,* with my *reedle, tootle ree.* Upon which the violin observed, in a very sprightly manner, *Niddle, niddle, niddle, niddle, niddle, niddle nee,* with my *nee,* with my long *nee;* which the bass-viol, in his gruff but sensible way, acknowledged to be as witty a thing as he had ever heard. This was followed by a general discourse, in which the violin took the lead, all the rest questioning and reasoning with one another, as hard as they could drive, to the admiration of the beholders, who were never tired of listening. They must have carried away a world of thoughts. For my part, my deafness came upon me. I never so much lamented it. There was a long story told by a hautboy, which was considered so admirable that the whole band fell into a transport of scratching and tooting. I observed the flute's mouth water, probably at some remarks on green peas, which had just come in season. It might have been guessed, by the gravity of the hearers, that the conversation chiefly ran upon the new king and queen; but I believe it was upon periwigs; for turning to that puppy Rawlinson, and asking what he concluded from all that, he had the face to tell me that it gave him a "heavenly satisfaction."

We laughed heartily at this sally against music.

Dr. Swift was very learned on the dessert. He said he owed his *fructification* to Sir William Temple. I observed that it was delightful to see so great a man as Sir William Temple so happy as he appears to have been. The *otium cum dignitate* is surely nowhere to be found, if not as he has painted it in his works.

DR. SWIFT. The *otium cum digging potatoes* is better. I could show you a dozen Irishmen (which is a great many for thriving ones) who have the advantage of him. Sir William was a great, but not a happy man. He had an ill stomach. What is worse, he gave me one. He taught me to eat platefuls of cherries and peaches, when I took no exercise.

A. H. What can one trust to, if the air of tranquillity in his writings is not to be depended on?

Mr. Pope. I believe he talks too much of his ease, to be considered very easy. It is an ill head that takes so much concern about its pillow.

Dr. Swift. Sir William Temple was a martyr to the "good sense" that came up in those days. He had sick blood, that required stirring; but because it was a high strain of good sense to agree with Epicurus and be of no religion, it was thought the highest possible strain, in anybody who should not go so far, to live in a garden as Epicurus did, and lie quiet, and be a philosopher. So Epicurus got a great stone in his kidneys; and Sir William used to be out of temper if his oranges got smutted.

I thought there was a little spleen in this account of Temple, which surprised me, considering old times. But if it be true that the giddiness, and even deafness, to which the Dean is subject, is owing to the philosopher's bad example, one can hardly wonder at its making him melancholy. He sat amid a heap of fruit without touching it.

Mr. Pope. Sir William, in his "Essay on Gardening," says he does not know how it is that Lucretius's account of the gods is thought more impious than Homer's, who makes them as full of bustle and bad passions as the meanest of us. Now, it is very clear : for the reason is, that Homer's gods have something in common with us, and are subject to our troubles and concerns ; whereas Lucretius's live like a parcel of *bons-vivants* by themselves, and care for nobody.

THE DEAN. There are two admirable good things in that essay. One is an old usurer's, who said that " no man could have peace of conscience that run out of his estate." The other is a Spanish proverb, that " a fool knows more in his own house than a wise man in another's."

The conversation turning upon our discussion last time respecting anglers, the Dean said he once asked a scrub who was fishing if he ever caught the fish called the Scream. The man protested he had never heard of such a fish. " What ! " says the Doctor, " you an angler, and never heard of the fish that gives a shriek when coming out of the water ? It is true it is not often found in these parts ; but ask any Crim Tartar, and he will tell you of it. 'Tis the only fish that has a voice ; and a sad, dismal sound it is." The man asked who could be so barbarous as to angle for a creature that shrieked ? " That," says the Doctor, " is another matter : but what do you think of fellows that I have seen, whose only reason for hooking and tearing all the fish they can get at, is that they do *not* scream ? " I shouted this *not* in his ear, and he almost shuffled himself into the river.

Mr. Walscott. Surely, Mr. Dean, this argument would strike the dullest.

Dr. Swift. Yes, if you could turn it into a box on the ear. Not else. They would fain give you one meantime, if they had the courage; for men have such a perverse dread of the very notion of doing wrong, that they would rather do it than be told of it. You know Mr. Wilcox of Hertfordshire? (to Mr. Pope). I once convinced him he did an inhuman thing in angling; at least, I must have gone very near to convince him; for he cut short the dispute by referring me to his friends for a good character. It gives one the spleen to see an honest man make such an owl of himself.

Mr. Pope. And all anglers, perhaps, as he was?

Dr. Swift. Very likely, 'faith. A parcel of sneaking, scoundrelly understandings get some honest man to do as they do, and then, forsooth, must dishonor him with the testimony of their good opinion. No: it requires a very rare benevolence, or as great an understanding, to see beyond even such a paltry thing as this angling, in angling times; about as much as it would take a good honest-hearted cannibal to see further than man-eating, or a goldsmith beyond his money. What! isn't Tow-wow a good husband and jaw-breaker; and must he not stand upon reputation?

Mr. Walscott. It is common to hear people among the lower orders talk of "the poor dumb animal," when they desire to rescue a cat or dog from ill treatment.

The Dean. Yes; and the cat is not dumb, nor the dog either. A horse is dumb; a fish is dumber; and I suppose this is the reason why the horse is the worst used of any creature, except trout and grayling. Come: this is melancholy talk. Mrs. Patty, why didn't you smoke the bull?

Mrs. Blount. Smoke the bull, sir?

Dr. Swift. Yes; I have just made a bull. I said horses were dumb, and fish dumber.

Mrs. Pope. Pray, Mr. Dean, why do they call those kind of mistakes *bulls?*

Dr. Swift. Why, madam, I can not tell; but I can tell you the prettiest bull that ever was made. An Irishman laid a wager with another, a bricklayer, that he could not carry him to the top of a building in his hod. The fellow took him up, and, at the risk of both their necks, landed him safely. "Well," cried the other, "you have done it; there's no denying that; but at the fourth story I had hopes."

Mr. Pope. Doctor, I believe you take the word *smoke* to be a modern cant phrase. I found it, when I was translating Homer, in old Chapman. He says that Juno "smoked" Ulysses through his disguise.

Mention was made of the strange version of Hobbes.

MR. POPE. You recollect, Mr. Honeycomb, the passage in the first book of Homer, where Apollo comes down to destroy the Greeks, and how his quiver sounded as he came?

"Yes, sir," said I, "very well": and I quoted from his translation:

"Fierce as he moved, the silver shafts resound."

MR. POPE. I was speaking of the original; but that line will do very well to contrast with Hobbes. What think you of

"His arrows chink as often as he jogs!"

Mr. Pope mentioned another passage just as ridiculous. I forget something of the first line, and a word in the second. Speaking of Jupiter, he says:

"With that —— his great black brow he nodded;
Wherewith (astonished) were the powers divine:
Olympus shook at shaking of his God-head,
And Thetis from it jumped into the brine."

MR. POPE. Dryden good-naturedly says of Hobbes, that he took to poetry when he was too old.

DEAN SWIFT (with an arch look). Perhaps had he begun at forty, as Dryden did, he would have been as great as my young master.

Mr. Walscott could not help laughing to hear Dryden, and at forty, called "my young master." However, he was going to say something, but desisted. I wish I could recollect many more things that were said, so as to do them justice. Altogether, the day was not quite so pleasant as the former one. With Mr. Pope, one is both tranquil and delighted. Dr. Swift somehow makes me restless. I could hear him talk all day long, but should like to be walking half the time, instead of sitting. Besides, he did not appear quite easy himself, notwithstanding what the boatman said; and he looked ill. I am told he is very anxious about the health of a friend in Ireland.

THE END.

www.ingramcontent.com/pod-product-compliance
Lightning Source LLC
Chambersburg PA
CBHW022009220426
43663CB00007B/1015